THE DAWN OF FOOD

Karen Mutton

Adventures Unlimited Press

The Dawn of Food

by Karen Mutton

ISBN: 978-1-948803-65-6

Published by:
Adventures Unlimited Press
One Adventure Place
Kempton, Illinois 60946 USA
auphq@frontiernet.net

Printed in the United States of America

AdventuresUnlimitedPress.com

Dedicated to John, who is always by my side.

THE DAWN OF FOOD

Karen Mutton

GLOSSARY

Agriculture—the cultivation of the ground, harvesting of crops, tillage, farming and animal husbandry

Apiculture—the raising of bees for their honey and wax

Aquaculture—rearing aquatic animals or cultivating aquatic plants for food

Archimedes screw—an early water pump attributed to Archimedes but used earlier in Mesopotamia cultures

Ayuveda—ancient Indian medical system

Ard—a simple plough consisting of a spike dragged through the soil

basin irrigation—irrigation of lad by surrounding it with embankments to form a basin which is then flooded

BCE—Before Common Era—before birth of Christ (BC)

BP—Before Present established as 1950

chinampa—a floating field on a shallow lake in ancient Mexico used to grow crops

cultigen—a plant whose origin is due to intentional human activity

cultivar—a hybridized plant

einkorn—an ancient heirloom wheat

emmer—one of the most ancient wheat species

fermentation—preservation by adding microbes

Fertile Crescent—a crescent-shaped region in the Middle East from the Levant to Tigis/Euphrates River where civilisation developed

Garum—Phoenician and Roman fish sauce

Latifundia—large landed estates in ancient Rome specialising in agriculture destined for export

Middens—ancient piles of domestic garbage

Miso– a Japanese fermented paste made from soybeans

Neolithic—the New Stone Age, birth of agriculture and pastoralism

Nixtamalization—preparation of corn in which it is soaked in limewater or another alkaline solution

Olynthus mill—ancient Greek hopper mill for grinding grain

Phytoliths—microscopic structures made of silica found in plant tissues

Puquios—ancient water systems in coastal desert regions of Peru and Chile

Qanat—ancient Persian water system with tunnels and vertical shafts

Quern—a stone hand mill for grinding grain

Saddle quern—a stone hand grinding mill with a large base stone and smaller upper stone, between which grain is placed and ground by a person

Sakia—an ancient wheel used to raise water

Subak irrigation—Balinese water system

Subsistence—the amount of food needed to stay alive

Swidden—a method of cutting and burning agriculture annually which is also called slash and burn

Symposium—in ancient Greece a gathering for men with music, food and drink

Teosinte—the ancestor of modern maize

Transhumance—the moving of cattle or other grazing animals to new pastures, often quite distant, according to the change in season

Trapetum—a stone olive press from ancient Rome

Viticulture—the science if growing, harvesting and processing grapes for wine

Wet-rice—growing rice in flooded fields

Yakchal – ancient Persian dome shaped structure which acts as an evaporative cooler

TABLES

Measurements provided in this book are either in imperial or metric depending on the original source. To find the equivalent, use the following table:

1 inch = 2.54 centimetres
1 foot = 0.30 metre
1 yard = 0.91 metre
1 mile = 1.69 kilometres
1 acre = 0.40 hectares
1 square mile = 2.59 square kilometres
1 metre = 0.062 miles
1 kilometres = 0.62 miles
1 hectare = 0.00386 square miles
1 acre foot = one acre of water at depth of one foot.

Some dating issues

BC (Before Christ) and AD (Anno Domini) have been replaced by BCE (Before Common Era) and CE (Common Era) by many historians as it removes the religious element. The two systems are interchangeable.

BP, Before Present is more confusing, as the Present is actually calculated from the year 1950, when radiocarbon dating (C14) was adopted. In 1950, atmospheric nuclear weapons testing dramatically increased the amount of radioactive carbon 14 in the atmosphere.

Archeologists often use BP as an imprecise estimate of an object's age. In this book, BP usually denotes older dates, while BCE and CE are more precise. It is important to remember that BP dates are at least 1,950 years older than BCE dates. The confusion arises from the original sources quoted.

TABLE OF CONTENTS

PART 1 THE NEOLITHIC REVOLUTION	**16**
NATUFIAN CULTURE	18
VINCA CULTURE	23
CROPS, ANIMALS AND FOOD PRODUCTION	25
FERMENTATION	37
POTTERY	42
PART 2 MESOPOTAMIA	**48**
SUMERIANS & AKKADIANS	50
BABYLONIANS	58
ASSYRIANS	61
NEO-BABYLONIANS	67
PERSIAN EMPIRE	67
HITTITES	71
THE LEVANT—PHOENICIANS	78
ISRAEL/PALESTINE	83
ARAB ISLAMIC EMPIRE	94
PART 3 AFRICA	**99**
EGYPT	100
WEST/SOUTHERN AFRICA	109

ETHIOPIA 110

CENTRAL & SUBSAHARAN AFRICA 111

PART 4 ANCIENT EUROPE 114

ANCIENT BRITONS & THE CELTS 115

SCANDINAVIA 120

ANCIENT GREECE- CRETE 123

MYCENAEANS 124

ARCHAIC GREEKS 125

CLASSICAL GREEKS 128

POTTERY 131

GREEK AGRICULTURAL MYTHOLOGY 133

HELLENISTIC EMPIRES 139

ANCIENT ITALY—THE ETRUSCANS 145

REPUBLICAN ROME 149

ROMAN EMPIRE 155

ROMAN FOOD TECHNOLOGIES 166

PART 5 ANCIENT ASIA AND OCEANIA 173

INDUS VALLEY 175

VEDIC INDIA 181

GUPTA INDIA 187

SRI LANKA 191

INDONESIA 186

SPICE ISLAND TRADE 199

THAILAND 202

VIETNAM 205

CAMBODIA 206

NEOLITHIC CHINA 212

BRONZE & IRON AGE AGRICULTURE IN CHINA 216

CHINESE CROPS & CUISINE 219

KOREA 232

JAPAN 235

PAPUA NEW GUINEA 241

LAPITA CULTURE 243

HAWAII 244

AUSTRALIA 247

PART 6 THE AMERICAS 257

MESOAMERICA—OLMECS & MAYANS 258

THE AZTECS 264

SOUTH AMERICA—ANDEAN HIGHLANDS 268

COASTAL REGIONS OF PERU 274

AMAZONIA 275

NORTH AMERICA 280

INTRODUCTION

Food Glorious Food!
We're anxious to try it
Three banquets a day
Our favorite diet
Just picture a great big steak
Fried, roasted, or stewed
Oh food! magical food! wonderful
food! marvelous food!

—Lyrics from "Oliver" musical

As 2023 progresses, hunger and starvation are stalking many areas of the globe due to climate change, the Ukraine war, the coronavirus pandemic, supply chain disruptions and the shortage of fertiliser. Before the Ukraine war which began in February 2022, the Horn of Africa countries of Sudan, South Sudan, Somalia, Ethiopia and Eritrea were already at risk of severe famine due to a prolonged drought. China and Europe also experienced extended summer droughts which threatened to dry up major rivers like the Rhine and Yangtze. On the other side of the climate spectrum, severe floods in Pakistan and Australia affected the food supply and the ability of these countries to grow and export food.

The Ukraine war and sanctions against Russia have also adversely affected the global supply of food, as these nations were top grain and fertiliser producers and exporters. Only one month into the conflict, the United Nations warned that 50 African and Middle Eastern countries were at immediate risk of famine.

> According to the United Nations International Fund for Agricultural Development (IFAD), the war in Ukraine has already caused food prices to rise and staple crop shortages in parts of Central Asia, the Middle East and North Africa.
>
> The Russian military operation in Ukraine severely reduced the number of shipments from the two countries,

which account for about 25% of global wheat exports, and 16% of global corn exports, which led to higher prices in global markets.

Since the beginning of the war, global wheat prices have jumped by 29.18% to exceed $11 a bushel, which is more than double the price of wheat in the same period last year when it ranged at $6 a bushel.

REFERENCE; "Teller Report," "50 Countries are threatened with starvation due to the Ukraine war.... And a 'hurricane of famines' may strike the world! March 18, 2022.
https://tinyurl.com/3mnnu6vd

Other war torn countries like Afghanistan, Yemen, Syria and Iraq, some of the earliest cradles of civilisation, have also experienced famine and food insecurity. In addition, islands like Madagascar, Comoros, Haiti and Timor Leste currently suffer from acute food shortages. The Ukraine war and sanctions against Russia have limited the export of natural gas to Europe, causing shortages and huge price increases in power, grain and fertiliser. Furthermore, intervention by governments pushing a radical climate change agenda have seen farmers forced off their land in the name of saving the environment by reducing carbon and nitrogen. The Dutch farmers, coerced into destroying their livestock, mounted a heroic challenge to the radical green agenda in the northern summer of 2022.

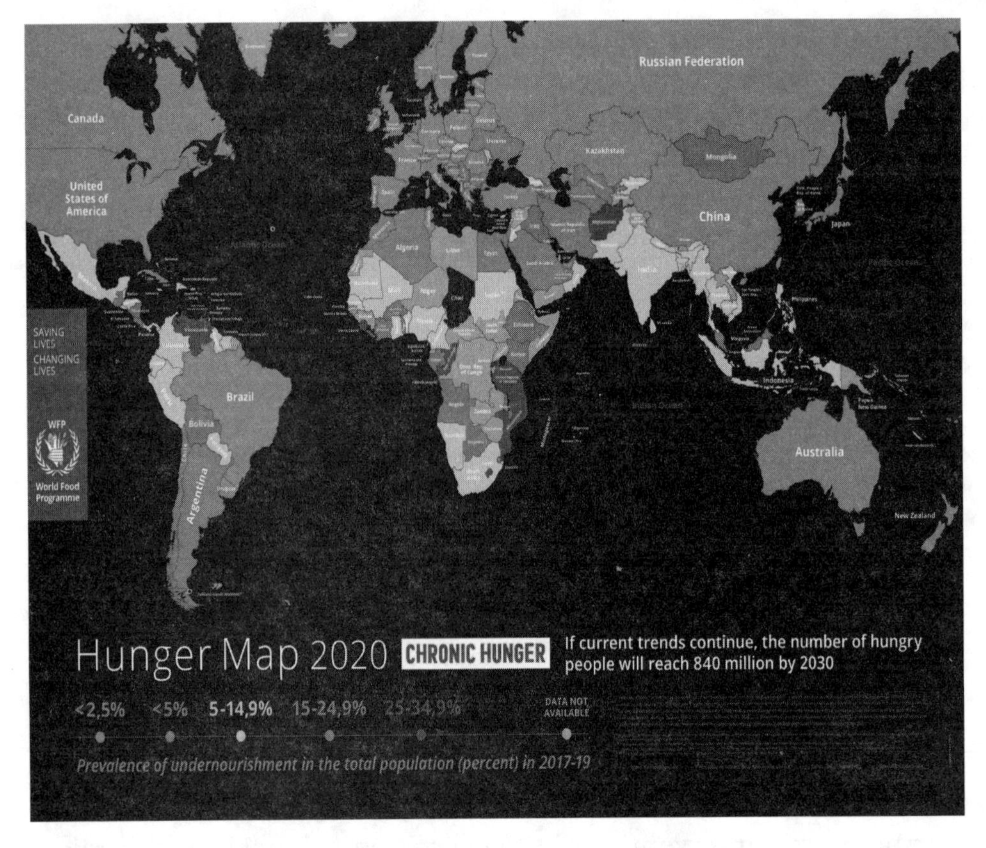

Image credit weforum.org

Droughts, floods and famines have always threatened humans who were trying to eke out an existence by farming crops and raising livestock for subsistence since the birth of agriculture over 12,000 years ago. How these farmers in countries around the world survived and thrived in often inhospitable environments is an inspirational story which needs to be told! While archeological journals are replete with academic papers discussing the origins of agriculture, the sheer wonderment of this process has yet to reach the popular imagination!

Over twelve thousand years ago, the Neolithic revolution in the Middle East, initiated the domestication of animals such as sheep, goats and cattle, and the cultivation of the "sacred grains" such as millet, barley and wheat in what was truly the most important revolution the world has ever known. This revolution also included the preservation of food by storage and the process of fermentation which used microbes; both requiring sedentary populations.

To the east, the rice paddies of central China, and Thailand were instrumental in the development of settlements and the trappings of civilisation. In the Americas, the cultivation of the "three sister" crops of maize, beans and squash brought about the first sedentary cultures.

The Neolithic Revolution in Europe was responsible for viticulture, fermented dairy products such as cheese, and animal husbandry. Great megalithic stone structures accompanied agriculture and sedentary societies.

Although humans have been cooking meat since the Palaeolithic Old Stone Age, the practice of deliberately cultivating crops extends back in time to:

- 12,500 years in the Levant region of the Middle East
- 10,500 years in China and Papua New Guinea
- 9,700 years in the Fertile Crescent
- 9,000 years in Mesoamerica
- 8,000 years in Greece
- 7,000 years in Egypt

The domestication of animals timeline:

- Wild sheep were managed in the Zagros Mountains (Turkey/Iran) 11,000 years ago and fully domesticated 8,000 years ago.
- Goats—Middle East 9,000 years ago
- Pigs—Middle East 11,000 years ago and China 8,500 years ago
- Cattle—8,000 years ago in the Middle East
- Chickens—China 7,000 years ago
- Alpacas and llamas—Andes 5,000 years ago

- Horses—Central Asia 4,000 years ago

The world is rediscovering the wonders of ancient food and long forgotten recipes which are being translated from old documents. The oldest recipes found inscribed on a 3,700-year-old cuneiform tablet come from the Akkadian Empire of ancient Mesopotamia, currently housed in the Yale Library. For the first time anyone can find ancient recipes at their fingertips on the internet.

Although some archeologists spend years meticulously studying pollen samples, digging into ancient middens and preparing statistical analyses in laboratories to increase our knowledge of ancient agriculture and cuisine, much research does not escape the dry pages of academic journals to reach the public. This book aims to cover six continents and the ancient cultures that have lived therein. Ancient agriculture, crops and farming methods are brought to life, along with recipes and cuisine, including;

- The oldest brewery (Egypt)
- Ancient irrigation in Mesopotamia, Persia and Peru
- The ancient spice trade
- The wonders of fermentation
- Bread, the staff of life
- The origin of citrus
- Agricultural gods and goddesses
- Food as medicine in China, Greece and Rome
- The wonderful fruits and vegetables from the Americas
- Ancient cookbooks in Rome and Arabia
- The oldest recipes in the world from Mesopotamia

PART 1

THE NEOLITHIC REVOLUTION

NATUFIAN CULTURE AND RECIPE

VINCA CULTURE

NEOLITHIC CROPS, ANIMALS, FOOD PRODUCTION & TECHNOLOGIES

FERMENTATION

NEOLITHIC POTTERY

OVENS, KILNS & GRANARIES

NATUFIANS in the Levant were the earliest farming culture in the Fertile Crescent.

The VINCA culture was one of the earliest European farming cultures.

Crops included emmer and einkorn wheat, barley, legumes, fruit and vegetables.

Domesticated animals were sheep, cows, pigs, goats and wildfowl.

FERMENTATION of grains for bread and beer, grapes, milk and vegetables began.

POTTERY was used for storage and transportation.

Ovens for cooking food and kilns for pottery were invented.

THE NEOLITHIC REVOLUTION

This term was first coined by Australian archeologist Gordon Childe in his 1936 book "Man Makes Himself" to describe the domestication of plants in the Middle East around 12,000 to 10,000 years ago. Prior to the adoption of agriculture, humans were living primarily as hunter/foragers, although evidence has emerged that some communities were harvesting and grinding wild grain well beyond 20,000 years ago.

Pre-Neolithic An Italian-led study of five ancient grindstones from 39,000 to 43,000 years ago shows that milling grains dates back to the transitional period between Neanderthals and *Homo sapiens.* Grindstones from Neanderthal sites show that Neanderthals were processing grains over one thousand years before modern humans, who later occupied the same site, were using pestles.

The authors concluded: "This study provides the oldest evidence of flour processing in Italian sites (Bombrini and Castelcivita) occupied by *H. sapiens* and the first direct evidence – starch grains in association with use-wear traces – of the use of this technology by Neanderthals in a crucial area for interrelations between the two human groups..."

Lippi, M et al, "New evidence of plant food processing in Italy before 40 ka," Quaternary Science Reviews,' July 15, 2023

https://www.sciencedirect.com/science/article/abs/pii/S0277379123002093?via%3Dihub

Archeological excavation at the Upper Palaeolithic site of Ohalo II in northern Israel has uncovered glossed flint blades for harvesting wild cereals 23.000 years ago. These sharp flint blades harvested near-ripe semi-green wild cereals before the grains were ripe.

A grinding stone dating to around 19,400 BP at Ohalo II indicates that the inhabitants processed the grain before consumption. These people living in Jordan, Syria and Israel (the Levant) around 14,000 years ago are called affluent foragers, because for the first time there were enough resources to allow them to live as foragers while settling down in towns such as Jericho.

THE NATUFIANS, the first settlers of Jericho, were affluent foragers from 15,000-11,500 years ago, before being amongst the earliest farmers in the world. Another site, Tell **Abu Hureyra** in Syria, was inhabited between 13,000 and 9,000 BP. During phase 1, these Natufians lived in villages and stored the food they hunted and foraged. Before the second phase, a climatic event called the Younger Dryas brought centuries of drought and cold to the area, which impacted the Natufians' ability to hunt and forage. They adapted by deliberately cultivating the cereal crop rye, and later emmer wheat and barley. When the protein supply of wild gazelle dried up, they domesticated sheep and goats. Natufians cooked meat on hearths and stored food in baskets. Shubayqa1, a 14,500-year-old site in Jordan shows evidence of Natufian flat bread making.

RECIPE- Natufian barley flour recreated with stone mortars, wooden pestles, sticks and sieves by Professor Mordechai Kislev's team at Bar-Ilan University.

Method: 1. Collection of spikelets, the coated grains of the cereal ear from wild barley. After ripening on the ground, the grains were separated from the stalks by beating against the threshing floor with a curved stick and sifting them through a large-holed sieve.

2. Conical mortars were used to transform wild grain into groats and flour. After the grains were beaten with a wooden pestle, the wider cones were used for removing the bristle that extends from the edge of the seed. Then the narrow cones, along with a wooden pestle, removed the grain husk.

3. After de-husking, the grain was scooped out of the conical mortar and placed into a small cup cut in the bedrock from where it was transferred for filtering in a small-gauge sieve.

REFERENCE: “Facts and Details,” “The Natufians” www.factsanddetails.com

The term **Neolithic** refers to the New Stone Age, when hunters/foragers began to grow their own crops, using farming implements made of stone and wood. This term is mainly specific to European and Middle Eastern cultures, as agriculture was developed at different times in specific areas. Agriculture sprang up in:

- Levant 12,000 BP (Before Present)
- Egypt around 9,000 BP
- China around 9,000 BP
- Papua New Guinea 10,000-6000 BP
- Crete, Europe 8500 BP
- Central Mexico 5,000-4,000 BP
- South America 5,000-4,000 BP

Why agriculture? Around 12,000 years ago the world was beginning to warm up after the last ice age, when the huge glacial ice sheets which covered much of Europe and North America began to melt. Archeologists have posited many theories as to why hunter/gatherers eventually settled down to more sedentary lives, and why this change occurred at different times across the globe, but climatic conditions must have been the deciding factor.

The Neolithic Revolution not only refers to the adoption of agriculture but also:

- Herding of animals for food, clothing and dairy
- Permanent settlements such as towns and cities
- Wide scale adoption of pottery for storage
- Emergence of trading networks
- Development of fermentation to preserve food

The domestication of plants began in the **Fertile Crescent**, spanning modern day Iraq, Syria, Lebanon, Palestine, Israel, Jordan, Egypt, south-eastern Turkey and western Iran. This area was home to the eight Neolithic founder crops, the wild progenitors of emmer wheat, einkorn, barley, flax, chickpea, pea, lentil and bitter vetch. Three of the most important species of domesticated animals; cows, goats and sheep also originated in the Fertile Crescent.

Cereal grinding stone Abu Hureyra, British Museum, Credit Zunkir CC BY-SA 4.0

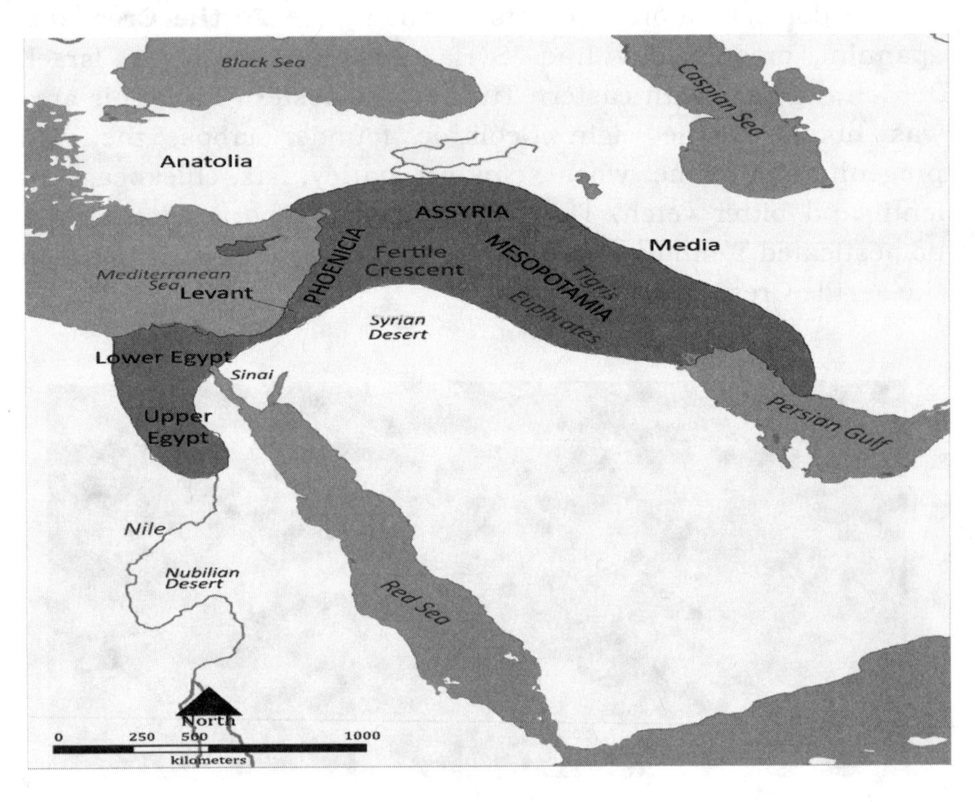

The Fertile Crescent, Credit Nafsadh CC BY-SA 4.0

Neolithic sites excavated in the Levant include Tell Aswad, which yielded domesticated emmer wheat, dated to 10,800 BP, as well as two-row hulled barley at Jericho and Wadi Faynan 16. Cultivation of wild plants has been found in Choga Gholan in Iran, dated to 12,000 BP on the eastern side of the Fertile Crescent.

The Anatolian town of **Catalhoyuk,** which was settled between 11,000 BP to 8400 BP, was first excavated by James Mellaart in 1958. In later levels of the site, the inhabitants were practising agriculture and the domestication of sheep and goats. Bins for storing cereals such as wheat and barley were found along with small female guardian statues. Peas were also grown, while almonds, pistachios and fruit were harvested from nearby trees. Sheep were already domesticated, and while remains of butchered bulls were discovered in several houses, it is unlikely that cattle domestication had begun.

Jericho became one of the oldest farming communities in the world when the Natufians adopted agriculture around ten thousand years ago. The inhabitants built a huge, thick wall around their houses where there were pits for cooking and stone querns for grinding flour. Not only did they build silos or granaries for storing surplus grain, but they also kept seeds and pulses in baskets and skins. Pottery was not invented until 8,000 years ago in this area.

Neolithic sites in Mesopotamia are split into four areas where farming was practised from the 8th millennium BP (6th millennium BCE.)

- **Hassunan** communities in the north dating from 6250-5300 BCE where there were small villages, rain fed agriculture and pottery, named after Tell Hassunan
- **Halafian** – 5500-4700 BCE, also in the north with rain fed agriculture, small hamlets and sophisticated pottery
- **Samarran** tradition in central Mesopotamia with irrigation agriculture, flax for clothing and trade, larger villages with T-shaped buildings which may have been storehouses for seed and grain. Sites excavated include Choga Mami and tell es-Sawwan. It is dated from 6500-6000 BCE.
- **Ubaid** tradition dating from 5800-4000 BCE in southern Iraq. Larger towns and cities like Eridu were built, irrigation of the alluvial soils

Neolithic culture spread to south-east Europe via Anatolia beginning around 8,500 years ago in Knossos, Crete, and Thessaly. Agriculture in the Balkans began a few centuries later. Around 9000 BP remains of domesticated plants and animals were found in Franchthi Cave, Greece, and it is generally accepted that the earliest Aegean farmers arrived from the Near East by boat.

The Linearbandkeramik culture (LBK) or Linear Pottery Ceramic culture was the first farming culture in central Europe, dated between 5400 and 4900 BCE. Some archeologists consider it to be the first Neolithic culture in Europe. LBK pottery consists of bowls made with clay and decorated with bands. Domesticated crops cultivated include emmer and einkorn wheat, crab apples, peas, lentils, flax, barley, linseed and poppies. They domesticated cattle, sheep, goats and pigs. The earliest LBK sites are found in the Hungarian plain and are dated to 5700 BCE. Later they spread to Germany, particularly the Rhineland, and Alsace by 5300 BCE.

The **VINCA CULTURE** occupied the Balkan area of south-eastern Europe, including Serbia, Kosovo, Romania Bulgaria Montenegro, North Macedonia and northern Greece during the Neolithic age from 5700-4500 BCE. It was a preliterate culture which introduced innovative agricultural techniques developed during the First Temperate Neolithic, to Europe. However, the Tartaria tablets, dated to 5500-5300 BCE, show pre-Sumerian symbols which may have been the earliest form of writing in the world. They were discovered at the Romanian site of Tartaria, discovered in 1961 by archeologist Nicolae Vlassa.

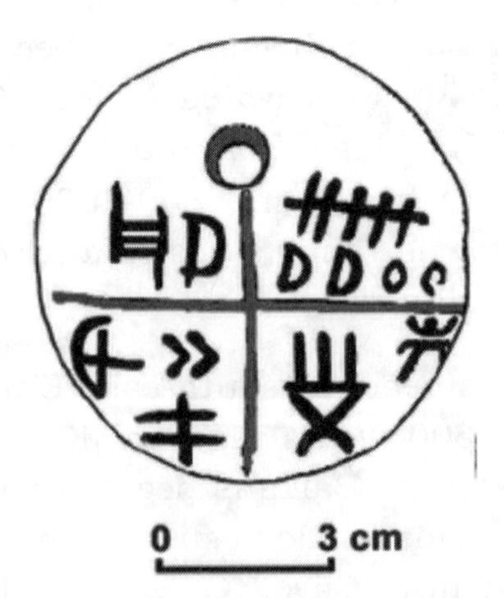

Tartaria tablet CC BY-SA 3.0

The Vinca residents practised a mixed subsistence economy of agriculture, animal husbandry, hunting and foraging. They introduced common wheat, oat and flax to middle Europe and concentrated upon barley. They increased crop yields and possibly used a cattle-driven plough.

Cattle were more important than sheep and goats and were kept not only for meat, but for milk, leather and as draft animals. Transhumant pastoralism, whereby livestock was herded from the lowland villages to the high pastures, occurred on a seasonal basis. The Vin□a also used wild food sources such as deer, boar, aurochs, fish, fowls, wild cereals, forest fruits and nuts.

The Vinca site of Plocnik has produced the earliest copper tools in the world, mined from sites like Rudna Glava. Lake Ohrid played a key role in the development of agriculture in Europe.

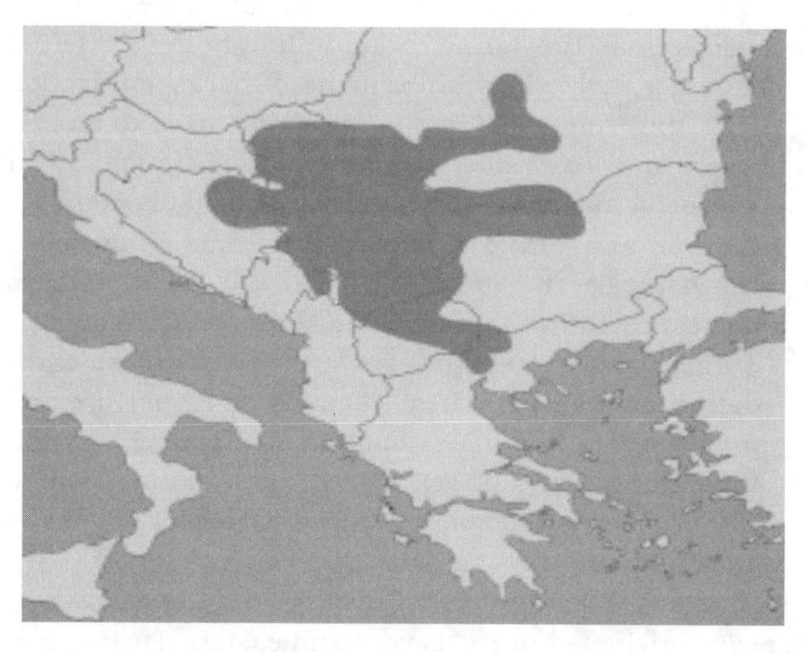

Vinca culture central Europe CC BY-SA 3.0

The settlement in the Bay of Ploca Micov Grad near the Macedonian town of Ohrid, now underwater, was excavated by archeologists from the University of Bern who uncovered preserved wooden piles of structures and other organic material. This 1.7 metre deep organic material is composed of harvested grain, wild plants and animals.

All Neolithic sites in Europe contain ceramics, as well as einkorn, emmer, barley, lentils, pigs, goats and cattle.

NEOLITHIC CROPS, ANIMALS AND FOOD PRODUCTION TECHNOLOGIES

- **Barley** *Hordeum vulgare* was one of the first cultivated grains, particularly in Eurasia around 10,000 years ago. Wild barley grew from Crete and North Africa in the west to Tibet in the east. The earliest evidence of consumption of wild barley comes from the site of Ohalo II, Israel, where grinding stones with traces of starch have been dated to 23,300 BP. The earliest evidence for domesticated barley comes from the Jarmo region in Iraq from 9,000-7,000 BP. Another early site with remains of barley cultivation is at Tell Abu Hureyra in Syria. Barley was used to make bread, beer and as fodder for animals. Throughout ancient Mesopotamia and Egypt, barley was a staple crop.
- **Emmer and einkorn wheat** were first cultivated in the southern Levant around 11,600 BP. Wild einkorn was domesticated in the Karacadag Mountains in south-eastern Turkey nearly 10,000 years ago. The earliest carbon 14 dating for einkorn wheat remains at Abu Hureyra range from 9800 to 9500 BP. Domesticated emmer was found in the earliest levels of Tell Aswad, Syria dating from 10,800 BP. Emmer cultivation reached Greece and the Indian subcontinent by 8500 BP, Egypt 8000 BP and Germany/Spain by 7000 BCE.
- **Rye** also originated in the Levant and eastern Turkey. Evidence uncovered at Tell Abu Hureya in Syria suggests that rye was cultivated around 13,000 years ago. Domesticated rye occurs in Neolithic sites such as Can Hasan III, Anatolia, but is absent

from the archeological record until 1800-1500 BCE in central Europe.

Temple of Horus Edfu, Egypt. Priest with wheat

- **Millet** is a crop which was initially domesticated in different parts of the world, especially East/South Asia and West/East Africa. In China, prosco and foxtail millet were cultivated at Cishan where phytoliths have been identified in storage pits dating from 10,300-8,700 years ago. Pit houses, pottery and stone tools like sickles were also uncovered. Millet noodles were discovered in a 4,000-year-old earthen ware bowl found at Lajia in north China. Korea and Korea domesticated millet between 5,500 and 6,000 years ago.

Little, kodo, barnyard and blackfinger millet were cultivated in India from 5,000 BP. In West Africa pearl millet was domesticated 5,500 years ago or even much earlier. The Sahel region was also growing pearl in 4,500 BCE. Europeans were harvesting wild millet 5,000 years ago in Greece and storing it in granaries centuries later in Macedonia.

- **Sorghum** has many varieties but *Sorghum bicolor* is native to Africa with numerous cultivated forms which go back to 7,000 years ago in the Sudan. It has always been cultivated as a food crop for humans and fodder for livestock. This sorghum is also known as durra, jowari or milo.
- **Legumes,** especially pulses like lentils, chick peas and have been cultivated in the Fertile Crescent since Neolithic times. Lentils are the oldest pulse crop and were cultivated at least 10,000 years ago in Iraq on the Euphrates River. Chickpeas are also known as grams, garbanzo beans and Egyptian peas. One of the earliest cultivated legumes, they have been discovered at Cayonu in Turkey (9900-9550 BP) and Jericho in the Levant. Chickpeas spread to the Mediterranean region around 8,000 BP and to India by 7,000 BP. A cave at L'Abeurador, Herault in southern France has yielded chickpeas dated to 8790 BP.
- **Soybeans** have been cultivated in China for at least 8,000 years, and were named one of the five sacred plants in China by Emperor Shennong in 2853 BCE, along with rice, wheat, barley and millet. Soya was fermented for sauce and processed for oil.
- **Grapes** were initially cultivated in the Middle East 6,000 to 8,000 years ago. From there grapes spread to the Aegean and Central Europe. Yeast, which occurs naturally on the skins of grapes, led to the discovery of fermentation for vinegar and wine. Traces of vinegar have been found in Egyptian urns from around 3000 BCE, while the Sumerians may have discovered it much earlier.
- **Olives** are native to the Mediterranean, Asia and Africa. The olive plant was first cultivated in the Mediterranean regions 7,000 years ago, and was grown commercially in Crete 5,000 years ago. The earliest evidence of olive oil production was found at an Early Chalcolithic site in Ein Zippori, Israel. Evidence from a small bowl found at Gerani Cave in Crete indicates that olive oil was produced at least 6,400 years ago. Another bowl dating from the Late

Neolithic contained a stew made of leafy vegetables and olive oil. It is believed that olive oil was produced in Neolithic times by placing olives in woven mats, and squeezed until the oil was collected in vats.

- **Nuts & seeds** grew prolifically in the Mediterranean and Middle East in ancient times and were prized for their nutritional value and oils.
- **Almonds** were eaten during the Palaeolithic era, but not domesticated until 3000 BCE in the Early Bronze Age.
- **Walnuts** were known throughout south-eastern Europe and all the way to the Himalayas. They have been unearthed in the Neanderthal site of Shanidar in Iraq and a Mesolithic dunghill in Switzerland. In ancient Persia only royalty ate walnuts, and cultivation began in Mesopotamia over 4,000 years ago.
- **Pinenuts** were consumed in the Mesolithic age, before the Neolithic, and in ancient Israel.
- **Pistashios** originated in Central Asia and north-west India. Remains of pistachios have been found in Iran and Afghanistan dating from 8,000-years-ago.
- **Sesame** originated in Africa and is one of the oldest oilseed crops. Charred remains of sesame discovered in India have been dated to 3500 BCE.
- **Salt** was mined and processed in Neolithic times in both Europe and China. Solnitsata in Bulgaria was originally a salt mine, providing the Balkans with salt since 5400 BCE. In Neolithic Romania, the Precucuteni people were boiling salt-laden spring water through the process of briquetage, a coarse ceramic material used to evaporate salt from brine or seawater at least 8,000 years ago. Bronze and Iron Age miners were extracting salt from deep mines in Hallstatt, Austria over three thousand years ago, and possibly much earlier.

The Neolithic Chinese were harvesting salt from the surface of Xiechi Lake in Shanxi at least 8,000 years ago. The Dawendou culture in northern China was also producing salt from underground brine deposits. In Anatolia salt may have been used for barter in connection with the obsidian trade.

Diorama of Neolithic miners, Hallstatt mine tour

Salt was not only necessary for dietary purposes, but it was also used to preserve food. The origins of salting food, particularly fish, have been lost in prehistory. Ancient cities such as Salzburg, Austria and Solnitsata were founded on salt mining and trade, as it was highly prized for its taste and preservative properties.

- Flax was first used for its fibres 30,000 years ago in Dzudzuana Cave, Georgia. Flax seeds used for oil have been found in Neolithic Tell Ramad in Syria and flax fabric fragments from Çatalhoyuk (9,000 years ago.) After being domesticated in the Fertile Crescent, its use spread to Switzerland, Germany, India and China 5,000 years ago.
- Cotton was domesticated and turned into fabric in several isolated cultures of the Old and New World. The oldest cotton fabric was uncovered in Huaca Prieta in Peru, dated to 6000 BCE. By 3000 BCE, cotton was being grown and processed throughout Mexico and Arizona. The early Indus Valley settlement of Mehrgarh was cultivating cotton 7,000 years ago, with large scale cultivation two thousand years

later. In Africa, cotton cloth was produced 7,000 years ago in eastern Sudan. The same tools such as looms, combs and spindles were invented independently in these areas.

Herding animals like sheep, goats and cattle live off pastures and have been domesticated in the Fertile Crescent and Mediterranean since Neolithic times.

Sheep and goats were the first mammals to be domesticated for meat, milk and skins around 11,000 years ago. Sheep, *Ovis aries* are believed to be descended from the Asiatic species of mouflon. They were selected for their lack of aggression, early sexual maturity, social nature and high reproduction rates. Archeological evidence from Iran suggests that the selection of woolly sheep may have begun

Sumerian ram in thicket, credit Jack 1956 CC BY-SA 3.0

around 8,000 years ago, and the earliest woven woollen garments have been dated to two thousand years later. The Neolithic Castelnovien people living near Marseilles in southern France were also keeping sheep 8,000 years ago.

By the Bronze Age, modern sheep were widespread throughout the Middle East, and featured as sacrifices in religions such as Judaism, and as a ram headed god in Egypt. Goats, too appeared in Bronze Age art from Sumerian times.

The wild bezoar ibex was the ancestor of the goat from the slopes of the Zagros and Taurus mountains in Iran, Iraq and Turkey around 11,000 years ago. They were also domesticated in Mehrgarh, Pakistan 9,000 years ago, central Anatolia, the Levant and China. Important archaeological sites with evidence for the initial process of goat domestication include Cayonu, Turkey (10,450 to 9950 BP), Tell Abu Hureyra, Syria (9950 to 9350 BP), Jericho, Israel (9450 BP), and Ain Ghazal, Jordan (9550 to 9450 BP.)

Domestic cattle, *Bos Taurus*, descended from a single herd of wild aurochs in the Near East about 10,500 years ago, according to a genetic study conducted in 2012. The *Bos indicus* was domesticated in the Indus Valley about 7,000 years ago. The earliest evidence for cattle domestication is the Pre-Pottery Neolithic cultures of the Taurus Mountains. Over the centuries, the decline in body size of the aurochs was demonstrated by archeological finds in Turkey, particularly Cayonu Tepesi. Taurine cattle were introduced into Europe about 8,500 years ago, and appeared in Asian sites (China, Mongolia and Korea) 5,000 years ago.

Evidence for the transition from wild Indian aurochs to *Bos indicus*, the humped zebu, 7,000 years ago has been discovered at the pre-Harappan site of Mehrgahr in the Indus Valley.

Oxen are male cattle which have been castrated in order to make them more docile. They have been used as draft animals for about

6,000 years. Their strength enabled them to drag sledges and later, ploughs and wheeled wagons in the Middle East and Europe. **Buffaloes** were first domesticated in Asia long before the Indus civilization. The water buffalo was bred to drag ploughs through flooded rice fields or haul carts on dry land across Asia.

Indus Valley seal with buffalo. CC0 1.0

Pigs were first domesticated from wild boar in the Near East Tigris River basin according to bones from sites like Cayonu. Remains of pigs have been dated to earlier than 11,400 years ago in Cyprus which must have come from the Near East. However, a study in 2015 indicated that pigs were domesticated separately in Western Asia and China, with Western Asian pigs eventually introduced into Europe where they were bred with wild boar around 8,500 years ago. The Chinese domesticated their pigs about 8,000 years ago and found them to be an invaluable source of protein.

In Papua New Guinea pigs have been domesticated for about 10,000 years and were probably introduced from the Molucca islands. Wild pigs were hunted for food, while domestic pigs were mostly kept for important ceremonial occasions, and only slaughtered when other sources of protein were unavailable.

Horses: Nomads of the Eurasian steppes from the Caucasus to Mongolia fermented mare's milk at least 6,000 years ago. Herodotus wrote of the Scythian method of creating mare's milk or kumis over 2,500 years ago in his famous "Histories."

> The Scythians blind all their slaves, to use them in preparing their milk. The plan they follow is to thrust tubes made of bone, not unlike our musical pipes, up the vulva of the mare, and then to blow into the tubes with their mouths, some milking while the others blow. They say that they do this because when the veins of the animal are full of air, the udder is forced down. The milk thus obtained is poured into deep wooden casks, about which the blind slaves are placed, and then the milk is stirred round. That which rises to the top is drawn off, and considered the best part; the under portion is of less account.-The Histories" Book 4

Kumis is similar to kefir, but produced from a liquid starter culture, whereas kefir is fermented milk using a colony of yeast and bacteria in the form of small gelatinous grains which were highly prized. Although the oldest traces were found in a 4,000-year-old tomb in Xiohe, China, the nomads of the Caucasus were fermenting kefir over 2,000 years ago

Chickens, ducks and geese

Domestic chickens, *Gallus domesticus* originated from red junglefowl in Thailand, South East Asia about 8,000 years ago. They spread to China and India 2,000 years later and bred with local species of wild junglefowl, forming distinct groups. Chickens appeared in the Middle East around 4,000 years ago, and reached Egypt for purposes of cockfighting about 1400 BCE. In about 800 BCE they were found in Europe, but not widely used for food until the Roman Empire.

By 7000 BCE, people in China and India were eating eggs, but chicken eggs were not generally eaten in Western Asia, Egypt or Europe until about 800 BCE. By 300 BCE, farmers in both Egypt and China learned to incubate eggs in warm clay ovens, so that the hens did not have to sit on them to hatch them.

Most **ducks** are descended from mallards which were first domesticated around 6,000 years ago in South East Asia. In ancient Egypt they were captured in nets and bred in captivity. Ducks were also farmed by the Romans and Malays for their eggs, meat and downy feathers.

Geese were domesticated in Europe, North Africa, Western Asia and Eastern Asia. The swan goose is the wild ancestor of Chinese geese. There is evidence geese were domesticated in Egypt more than 4,000 years ago and possibly 5,000 years ago in Southern Europe. They were also herded by ancient Mesopotamians for food and sacrifices.

Beekeeping, or apiculture, the maintenance of bee colonies with man-made hives, began in the Neolithic around 9,000 years ago. The oldest depiction of bees comes from a cave painting in Spain dating from 8,000 BP. Honey and beeswax in pottery vessels from that time were found in North Africa, and depicted in Egyptian art 4,500 years ago showing how smoke was used to subdue the bee hive before the honey could be collected and stored in jars. In ancient Crete, hives, smoking pots, honey extractors and other beekeeping paraphernalia were found in excavations at Knossos.

Spanish rock art of bees, valencia-international.com

Potsherds throughout the Middle East show evidence of beeswax, which was also important for medicinal and industrial purposes.

The following inscription from *Suhum* in Mesopotamia describes bees.

> I am Shamash-resh-uṣur, the governor of Suhu and the land of Mari. Bees that collect honey, which none of my ancestors had ever seen or brought into the land of Suhu, I brought down from the mountain of the men of Habha, and made them settle in the orchards of the town 'Gabbari-built-it'. They collect honey and wax, and I know how to melt the honey and wax – and the gardeners know too. Whoever comes in the future, may he ask the old men of the town, (who will say) thus: "They are the buildings of Shamash-

resh-uşur, the governor of Suhu, who introduced honey bees into the land of Suhu."

Translated text from Babylonian stele, (Dalley, *2002)* Text and image of this stele from https://kids.kiddle.co/Beekeeping

In China, the earliest written records, the oracle bones dated to 1100 BCE, called bees which were living in bee colonies raised by farmers *mi-feng*. A few centuries later there were records of beehives, techniques to domesticate wild bee colonies and a thriving market for beeswax and honey. Eventually, the Chinese used beeswax to make candles long before the technique was known in Europe.

Mead, or fermented honey wine is dated to 7000 BCE in China and was produced in prehistoric times throughout Europe. The Bell Beaker Culture of Europe, dating from 2800-1800 BCE, also produced mead. Roman naturalist Columella published the following recipe for mead in his *De re rustica in* 60 CE.

"Take rainwater kept for several years, and mix a sextarius of this water with a [Roman] pound of honey. For a weaker mead, mix a sextarius of water with nine ounces of honey. The whole is exposed to the sun for 40 days, and then left on a shelf near the fire. If you have no rain water, then boil spring water."

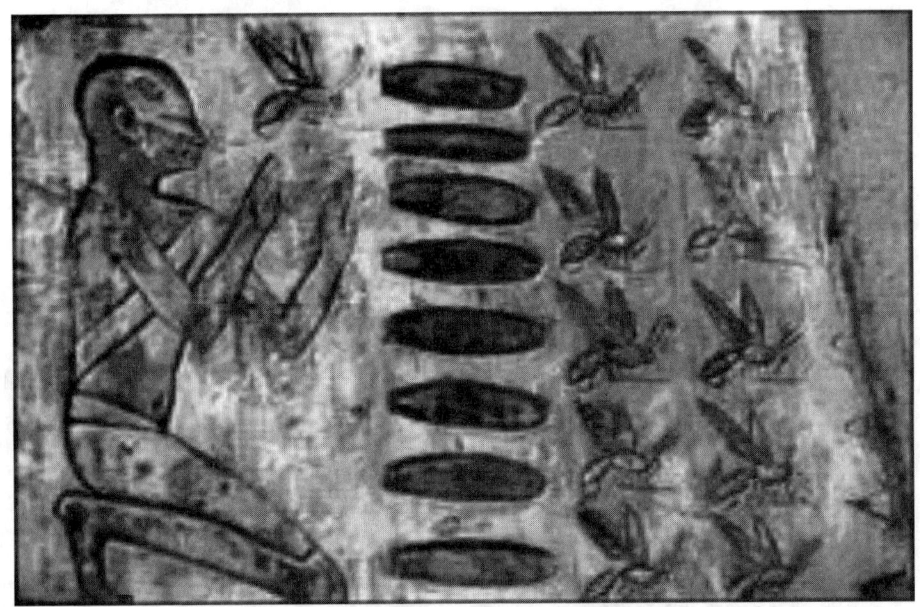

Bee-keeping in Egypt

FERMENTATION, the chemical breakdown of a substance by bacteria, yeasts or other micro-organisms, was the most important technical advance in food processing in the Neolithic age. Fermentation is responsible for the following foods and beverages: beer, wine, vinegar, mead, cheese, yoghurt, kefir, bread, sauerkraut, kimchi, miso, natto, soy sauce and pickled products.

Beer is one of the oldest fermented drinks in the world and probably originated in Mesopotamia or Iran. Chemical tests of ancient pottery jars reveal that beer in Iran was produced 7,000 years ago, while a 6,000-year-old tablet in Mesopotamia depicts people consuming a beverage, possibly beer, through reed straws from a communal bowl. The Sumerian beer goddess, Ninkasi appears in the Hymn to Ninkasi which is a recipe for brewing beer from barley bread.

> When you pour out the filtered beer of the collector vat,
> It is [like] the onrush of Tigris and Euphrates.
> Ninkasi, you are the one who pours out the filtered beer of the collector vat,
> It is [like] the onrush of Tigris and Euphrates.

Prince, J. Dyneley, (1916) "A Hymn to Ninkasi," 'The American Journal of Semitic Languages and Literatures." 33

Sumerian tablet of people using reed straws

The earliest chemically confirmed barley beer was discovered at Godin Tepe, in the Zagros Mountains, Iran, where fragments of a jug, over 5,000 years old, were found to be coated with beerstone, a by-product of the brewing process. Beer was also brewed in Neolithic Europe at the same time. In China, fermented alcoholic drinks were created as far back as 9,000 years ago, with similar production methods to those in ancient Egypt and Mesopotamia.

Egyptian beer was made from baked barley bread and also used in religious practices over 5,000 years ago. In February 2021, archeologists unearthed a 5,000- year-old brewery in Abydos, dating back to the Early Dynastic King Narmer. This large complex was comprised of eight large sections for beer production, each containing 40 clay pots used to warm the grain and water mixtures. The brewery was capable of producing as much as 22.400 litres (5,900 gallons) of beer at a time.

A 2,550-year-old Celtic settlement at Eberdingen-Hochdorf in Germany had six specially constructed ditches where barley malt, a key beer ingredient, was produced. According to archeologist Hans-Peter Stika, of the University of Hohenheim in Stuttgart, thousands of charred barley grains discovered in the ditches, came from a large malt-making enterprisc.

In historical European times, beer was less favoured than wine, except by northerners such as Thracians and Germanics. Nevertheless, the Greeks were fond of barley wine and Greek philosopher Sophocles felt the need to caution moderation when consuming beer. Hops were not used in beer production until the 9th century in France.

Cider is made from the fermented juice of apples. The British Celts were producing it from crab apples and it quickly spread throughout the Roman Empire after the invasion of Britain in 55 BCE.

Wine making can be traced back to 7000 BCE in Jiahu, China, where residues on pottery sherds come from a mixture of rice, honey and fruit. Native grape seeds have also been found at this site. The earliest evidence of grape processing in Western Asia include a site at Lake Zeriber, Iran where grape pollen was found in a soil core dated to 4300 BCE. Excavations at a Greek site called Dikili Tash have revealed grape pips and empty skins dated to between 4400-4000 BCE, the earliest so far discovered in the Aegean. Viticulture was also very important in Minoan and Mycenaean cultures which followed the Neolithic in Greece.

The oldest known winery was discovered in the "Areni-1" cave in Vayots Dzor, Armenia where a wine press, fermentation vats, jars and cups have been dated to 4000 BCE. This winery consisted of a trough measuring 3 x 3.5 feet with a drain leading to a 2-foot-long vat that could contain 15 gallons (57 l) of wine.

The seeds were from *Vitis vinifera*, the grape still used to make wine.

In Egypt, a thriving royal winemaking industry was established in the Nile Delta following the introduction of grape cultivation from the Levant in about 3000 BCE. Shedeh, red wine was the most precious drink in the New Kingdom because of its resemblance to blood and many blood myths. White wine was also available during the time of New Kingdom pharaoh Tutankhamun.

Egyptian wine production

In Neolithic times and well into Egyptian times, grapes were pressed in large troughs or vats by foot. It seems likely that the first wine presses, using planks of wood to extract the juice from grapes, may have been used in ancient Crete, along with the foot stomping method.

Bread making first appeared as flatbread made by Natufian hunter-gatherers from wild wheat, barley and roots between

14,600-11,600 years ago in the Fertile Crescent. The archeological site of Shubayqa 1 in Jordan predated the earliest Neolithic bread making sites in Turkey and Europe by thousands of years.

To create flour and bread, new technologies were invented, such as the sickle, mortar and pestle, sieve and oven. Neolithic bread was unleavened; the ancient Egyptians later discovered how to leaven bread using yeast. Leavened bread uses fermented dough, left over from the previous day's baking. This process is also known as sourdough.

Porridge, also known as gruel, which is made by boiling ground grains in water or milk, has been around since the Stone Age in many parts of the world. Although traditionally associated with oats and Scotland, it can be made with wheat, rye, spelt, barley, millet, sorghum or maize. In Neolithic times porridge could be made by adding hot stones to earthenware vessels containing the grain until the water boiled, or cooked on heated stone slabs. Groats are the hulled kernels of cereal grains such as oat, wheat, rye and barley. They are wholegrains that include the cereal germ and bran portion of the grain.

Cheese was originally made from sheep milk during the Neolithic in Europe and the Middle East around 8,000 years ago. It may have been discovered accidently by the practice of storing milk in containers made from the stomachs of animals. Rennet is a natural enzyme found in the stomach of ruminant animals which causes milk to coagulate, separating the curds and whey. Or Neolithic people could have salted curdled milk for preservation purposes.

Many Neolithic humans were lactose intolerant as adults, so early milk production may have been for suckling infants, or to create lactose reduced products like butter and cheese. The first cheeses were probably soft like feta or Indian paneer.

By 7,000 years ago, cow's milk was used to make cheese in Poland, according to studies of perforated ceramic material that had been used as sieves to separate the curds from whey. Traces of 7,000-year-old cheese have also been detected on Croatian pottery. Milk residue has also been discovered on 8,000-year-old ceramic vessels from Çatalhoyuk in Anatolia.

In historical times, cheese was made in Sumer, Hittite Anatolia, Greece, Rome, India, Celtic Europe and Iran.

Yogurt is a dairy product made by heating milk to a high temperature and adding in healthy bacteria to metabolise the milk's sugars. It originated in the Near East during the Neolithic period and probably began as sour or curdled milk in those hot climates. Indians also began fermenting milk around 8,000-6,000 years ago. With cows a sacred animal in Hindu India, yogurt and dairy have been long prized on the sub-continent. Across Eastern and southern Europe, yogurt was known, especially in Bulgaria where milk stored in animal stomachs created a warm environment for bacteria to grow and transform milk into yogurt or cheese.

Vinegar has a long history extending into antiquity and probably predates wine. Around 5,000 years ago, the Babylonians and Egyptians used the fruit and sap of the date palm which fermented naturally into vinegar after contact with air. The Greeks and Romans made vinegars from wine, using it for culinary and health purposes.

NEOLITHIC POTTERY is comprised of three main types of ceramic ware:

1. Earthenware which is the simplest and oldest pottery fired at the lowest setting between 1000 and 1200 degrees Celsius.
2. Stoneware is thicker, higher temperature fired pottery at 1100-1300 degrees Celsius and often covered with a powdered glass glaze and then re-fired at a higher temperature. The glaze makes an impenetrable surface.

3. Porcelain is the finest variation and produces a ringing tone when tapped. It was perfected in China.

Pottery is a necessity in sedentary, farming societies for conveying liquids like wine or oil over long distances, such as trade, for storage and for cooking vessels. The following refers to Middle Eastern pottery styles:

1. Between 7500-5800 BCE, pots were burnished or monochrome. Some were painted and of exceptional quality, such as Samarran ceramics in Mesopotamia.
2. The Middle Neolithic period known as "Sesklo Culture" pottery designs were variations of zigzag lines.
3. The Late Neolithic era, also known as "Dimini Culture" included a large range of pottery styles with paint and decoration on a red burnished backdrop.
4. From 4500-3300 BCE the Final Neolithic phase, also known as the "Rachmani Phase," included lugs on vases and thick paste instead of paint as embellishment.

Examples of Neolithic pottery from around the world:

- The earliest ceramics in the world have been discovered in Jomon sites dating to 14,450 BCE at the Odai Yamamoto 1 site in Japan. Other extremely old ceramics have been found in Kyushu, dating from 12,700 BCE. All Jomon pots were made by hand and had shells, mica and other elements mixed in, as the potter's wheel had not been invented. The pots were dried and fired in a huge fire at around 900 degrees Celsius. These ceramics included containers, bowls, dishes and jugs with spouts.
- Earthenware pottery was manufactured in China around 10,000 years ago, some with sophisticated designs.
- Ceramic sherds dating from 9500 BCE have been found in Sub-Saharan Mali.

- In Egypt, faience, earthenware pottery covered with a glaze was preferred. The earliest faience studio, furnished with lined block ovens, was found in Abydos Egypt dated to 5500 BCE. Precious stones were added with other minerals to make Egyptian faience.
- In the Neolithic Greek Aegean islands, jars were created around 3000 BCE.
- Earthenware was utilised in India at Mehrgarh from 5500-4800 BCE.
- The earliest pots appeared Britain around 4000 BCE and are known as Carinated Bowls. The best known Neolithic pottery is Grooved ware from the island of Orkney.
- The Linear Pottery culture (LBK) of Europe flourished from 5500-4500 BCE. Important sites have been found along the Danube, Elbe and Rhine. The pottery consists of cups, bowls, vases and jugs.

REFERENCE: "Neolithic Pottery History: 3 Types of Neolithic Pottery," "Crafts Hero'
https://craftshero.com/neolithic/pottery

Japan Jomon 10,000 BP

Kilns and ovens

The earliest kilns were pits dug into the ground with combustible materials and pottery stacked on top. After cooling, the pots were cleaned and used. Due to the low temperatures of

1000-1200 degrees F, the pots were porous and fragile, as glazing was not possible.

The world's oldest kiln has been traced to Mesopotamia and is 8,000 years old. Another kiln of that age was located in Yarim Tepe, Iraq, which had more than 1,500 rectangular furnaces and ceramic ovens used for cooking. Neolithic kilns in Bulgaria and Romania dating from 7,000 years ago, most likely contained two chambers, the pit and a clay beehive shaped firing chamber raised above the ground.

The Early Neolithic site of Portonovo-Fosso, in central Italy yielded 23 ovens but an absence of dwellings, and is dated to the 6th millennium BCE. The ovens have a circular base up to 2 metres in diameter and a height of about 0.50 metres as well as a single central opening faced on a large pit. Each structure had a lining, variable in thickness. Some of the ovens were used for funerary purposes, while others were for cooking or firing pottery. It is likely that the temperature did not exceed 500 degrees Celsius.

A 6,500-year-old oven was unearthed in a Neolithic site at Bapska in Croatia which was used to cook food, hot water and central heating for the home. It was permanently heated all day long and probably also was useful for baking bread.

Granaries for storing grain were built in Neolithic times. The earliest granaries were discovered in the Jordan Valley and date back to around 11,000 years ago. They measured about 10 x 10 feet square and featured raised floors which allowed the air to circulate and provide protection from vermin like mice. According to scientists, these granaries precede the emergence of fully domesticated plants by at least 1,000 years, and present a critical evolutionary shift in the relationship between people and plant foods.

Egypt's earliest granaries containing remnants of wheat and barley were found at Fayum dating from 5200 BCE by Gertrude

Caton-Thompson and Elinor Gardner in 1929. The basketry linings of the pits were in such a good state of preservation that they were able to transport them to the UK for study. Apart from 3.4 kg of grain, the pits yielded a wooden sickle with a flint blade.

REFERENCE: Wendrich, W, Cappers, R, “Egypt’s earliest granaries: evidence from the Fayum,” “Egyptian Archeology,” 2005

https://www.academia.edu/866611/Egypts_earliest_granaries_evidence_from_the_Fayum

PART 2

THE NEAR EAST

MESOPOTAMIA

SUMERIANS & AKKADIANS

BABYLONIANS

ASSYRIANS

NEO BABYLONIANS

PERSIANS

HITTITES

PHOENICIANS

THE ISRAELITES

THE ARAB ISLAMIC EMPIRE

IRRIGATION began in Sumer and spread across Mesopotamia and the Levant.

AKKADIAN tablets describe the oldest recipes in the world.

PERSIANS invented the ingenious qanat irrigation system and introduced many Central Asian foods to the Near East.

MEDIEVAL ARABS wrote comprehensive cookbooks which still survive

MESOPOTAMIA

Mesopotamia is known as the “land between two rivers,” the Tigris and Euphrates in Iraq, Western Asia. It is widely regarded as the cradle of civilisation, where great city states such as Sumer, Akkad, Babylonia and Assyria rose and fell over thousands of years. The area to the south was first urbanised by the Sumerians, whose brilliant inventions included writing, time keeping, cities, sailboats, the wheel and canals.

Mesopotamia is also part of the “Fertile Crescent” due to the rich alluvial soils deposited annually by the flooding rivers. However, the rivers sometimes did not flood, which was disastrous for food production and water management. The **Samarran** culture in northern Mesopotamia practised irrigation at towns such as Tell es-Sawwan from 5500-4800 BCE, while the oldest irrigation canal was in operation at Choga Mami around 6000 BCE. Irrigation supported livestock such as cattle, sheep and goats as well as wheat, barley and flax. This culture was also known for its finely made pottery.

From the 5th millennium BCE, the **Ubaidian** culture in the southern marshland gulf region drained the swamps for agriculture and established industries such as pottery, weaving and masonry. Tools like sickles were often made of hard fired clay, unlike those in the north which were made of stone or metal.

SUMERIANS

Around 4000 BCE, the southern inhabitants built a large network of canals for irrigation and transportation of water craft such as the sailboat. This was known as the **Uruk period** and lasted until 2900 BCE, a period of time in which copper and bronze were used and large cities developed. Southern Mesopotamia was productive because of the irrigation system, and focused on the cultivation of barley, date palms and other legumes/fruits, as well as the pasturing of sheep.

According to Nasrat Aam and Nadhir Al-Ansari in their paper, 'The Sumerians and the Akkadians: The Forerunners of the First Civilisation 2900-2003 BC"

"It is an established fact that the first successful efforts to control the flow of water on a very large scale were made in Mesopotamia. The Sumerians in southern Mesopotamia built city walls and temples and dug canals, which may be counted as some of the earlier of the world's first engineering works of their kind. It is also of interest to note that these people from the beginning of recorded history fought over water rights and agricultural land, and irrigation were (sic) extremely vital to them. Flooding problems were more serious in here than in Egypt because the Tigris and Euphrates were much swifter than the Nile and carried several times more silt per unit volume of water than the Nile did. This resulted in rivers rising faster and changing their courses more often in Mesopotamia.

The Sumerians had to solve much bigger hydraulic problems than the Egyptians whose civilization had not yet developed at that time yet. The processes leading to the Sumerian Civilisation cannot be understood except as creative adaptation to the priceless resources of the Tigris and Euphrates waters which led to this civilisation during the third millennia BC."

https://www.researchgate.net/publication/339782641_The_Sumerians_and_the_Akkadians_The_Forerunners_of_the_First_Civilization_2900-2003BC

The twin rivers often spilled their floodwaters over the banks which eventually raised the rivers to levels higher than the adjacent land, allowing gravity irrigation to flood the fields without the need of water lifting devices such as the shaduf. The Sumerians learned to control the floods by building canals, dykes, levees, weirs and reservoirs. One canal reached 120-metres-wide, which was large enough to permit navigation, whilst a smaller canal was 190 metres long 1 metre wide and 0.25 m deep. Weirs consisted of two gates made of wood or reeds and bitumen, while more sophisticated hydraulic structures were built with mud brick.

Maintenance of the canals was ongoing, with large gangs of workers required to free the canals of mud and silt. Major canals were supervised by high officials who reported to the king. Secondary canals were owned and controlled by the farmers and plot owners who had to clean them. In 2200 BCE the Shatt-el-hai canal linked the Tigris and Euphrates rivers.

Farming Tools

Sumerians were great innovators who invented such farming tools as the Ard, which was an oxen pulled wooden plough, and even a seed sowing machine that could plant seeds more quickly than sowing by hand, by utilising oxen power. The **ard,** or scratch plough, is a simple light plough without a mouldboard. Oxen, also known as bullocks, were castrated male bovines which were bred as draft animals. The invention of the wheel improved farming methods with the use of carts pulled by oxen.

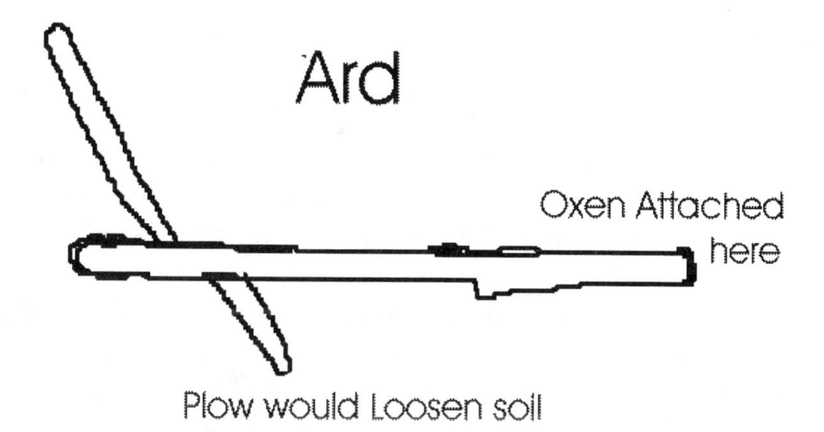

CROPS

By 5000BCE, the Sumerians had invented or utilised a large number of core agricultural techniques such as irrigation, mono-cropping involving the use of the plough, large scale intensive cultivation and a specialised labour force under bureaucratic control.

A 100-line Sumerian text known as the “Farmer’s Almanac” gave written instructions for the cultivation of barley.

1. After the spring equinox towards the end of summer, the farmers would flood their fields and drain them to loosen up the baked soil after the summer heat.

2. At the beginning of autumn, soil preparation would begin. The oxen would stomp the ground to kill weeds before the farmers used them to drag the fields with the ard plough drawn by four oxen. After, the ground was ploughed, harrowed, raked and pulverized with a mattock before planting the seed. The work could be completed with a hoe and a spade.
3. Sowing took place in autumn. The ploughs were equipped with a seeder, a funnel designed to leave the seed buried beneath the plough as it turned the soil.
4. Harvest was in spring just before the river rose, with teams consisting of a reaper, binder and sheaf handler. Three methods were used to separate the cereal heads from the stalks:

a) Threshing wagons, driven by oxen

b) Hand held sickles

c) Threshing boards consisted of a wooden board pulled by oxen with flints attached to separate the grain from the stems. Finally they winnowed the grain to separate the chaff.

Postgate, J. N, "Early Mesopotamia, Society and Economy at the Dawn of History," 1992, London and New York.

Apart from the main crop of barley, the Sumerians grew emmer wheat, chickpeas, lentils, dates, onions, garlic, lettuce, sesame and mustard as well as numerous fruits such as dates, apples, pomegranates and plums.

Food

For years archeologists believed that the Sumerians mainly ate a diet of pottage, probably legumes mixed with barley, along with barley bread and beer. Quite recently, French archeologist Jean Bottero decoded an Akkadian cuneiform tablet dating from 1900 BCE which contained a Sumerian-Akkadian dictionary with lists of over 800 food items, including 20 different cheeses and 300 bread recipes.

We now know that their diet was varied. Lamb, pork and 50 species of fish were favourites with urban Sumerians. Goats were prized for both their meat and milk, as were ewes and cows. Barley was the main staple grain of the diet which made bread, cakes and paste. Barley and vegetables were often made into soup or enriched bread. Fruits such as apples, figs, dates, grapes and pomegranates were also grown. Some were dried or preserved with honey which was added to many types of food. Meat was preserved with salt. Geese and ducks provided meat and eggs. Barley ale and beer were also consumed.

Cooking methods included roasting, boiling, barbequing or broiling. Preserving methods were smoking, drying, fermenting and salting. Street vendors produced many types of fast food including fried fish, grilled goat, mutton and pork. Bread was coarse, flat and unleavened, often accompanied with butter, milk and cheese. The Sumerians themselves wrote about the origins of their food in their literature.

The story "How Grain Came from Sumer":

> "Men used to eat grass with their mouths like sheep. In those times, they did not know grain, barley or flax. An brought these down from the interior of heaven. Enlil lifted his gaze around as a stag lifts its horns when climbing the terraced (...) hills. He looked southwards and saw the wide sea; he looked northwards and saw the mountain of aromatic cedars. Enlil piled up the barley, gave it to the mountain. He piled up the bounty of the Land, gave the innuha barley to the mountain. He closed off access to the wide-open hill. He (...) its lock, which heaven and earth shut fast, its bolt, which (...)
>
> "Then Ninazu (...) and said to his brother Ninmada: "Let us go to the mountain, to the mountain where barley and flax grow; (...) the rolling river, where the water wells up from the earth. Let us fetch the barley down from its mountain, let us

introduce the innuha barley into Sumer. Let us make barley known in Sumer, which knows no barley."

Ninmada, the worshipper of An, replied to him: "Since our father has not given the command, since Enlil has not given the command, how can we go there to the mountain? How can we bring down the barley from its mountain? How can we introduce the innuha grain into Sumer? How can we make barley known in Sumer, which knows no barley? "Come, let us go to Utu of heaven, who as he lies there, as he lies there, sleeps a sound sleep, to the hero, the son of Ningal, who as he lies there sleeps a sound sleep." He raised his hands towards Utu of the seventy doors." Black, J.A. et al, 1998, 1999, 2000, Electronic Text Corpus of Sumerian Literature, Oxford University, piney.com

The Sumerians brewed beer from grains as early as 3500 BCE. While the oldest known recipe for beer was contained in the 3,900-year-old poem "Hymn to Ninkasi," the consumption of beer was enjoyed by the wild man Enkidu in the earlier "Epic of Gilgamesh."

Land ownership

There were no written records prior to the Uruk period, but the construction of irrigation canals by the Samarran and Ubaid cultures indicate that there were social hierarchies and centralised administration. Later, kingship developed and nobles became the biggest landowners, along with the temples.

As the city of Uruk expanded, a huge enclosing wall was erected, and a radius of about 8 km of land around the city was required to provide sufficient food for the inhabitants. By 3000 BCE, Uruk had grown to 200 hectares and was surrounded by four settlements of about 40 hectares. A temple economy was established, with large amounts of food required to offer to the deities and then feed the temple workers who may have numbered up to 3,000.

Temple land could not be sold or alienated and was of three types:
1. Nigenna property was used for the maintenance of the temple.
2. Kurra was land dedicated to the temple priests and workers.
3. Urulal was land given to others in exchange for other land.

Temples owned large tracts of land that could be used for agriculture or grazing herds of cattle, sheep or goats. The animals were bred for food, wool or sacrifice. There was still some private land ownership where peasants owned houses, land and ponds, but in later years, the king and nobles were the main landowners.

The **Akkadians** were Semitic nomads from the Arabian Peninsula who subjugated the Sumerians and built the first empire in Mesopotamia from 2234-2154 BCE. Their sphere of influence included the Levant, Iran and Anatolia. After conquering the Sumerian kings they used the cuneiform alphabet for their own language. There is evidence of a huge drought around 2200 BCE which impacted agriculture, and clashes with the pastoralist nomadic Amorities over water.

The Akkadians, under kings such as Sargon, utilised the irrigated farmlands of the south and the rain-fed areas of northern Iraq.

The world's oldest known culinary recipes come from ancient Akkad in the form of a group of clay tablets kept in the Yale Babylonian Collection. These dishes include breads, porridges, stews, soups, cakes, pies and roasts. Many dishes were served together and continuously during a seating. Sweet and savoury dishes were apparently served together.

The following Akkadian recipe for beef broth was published in the course notes of the Great Courses lectures on 'Food; A

Culinary History' by Professor Ken Albala. Although the Akkadians conquered the Sumerians around 2000 BCE, they carried on most of their cultural traditions.

Recipe: Tuh'u Beet Broth

(adapted from Jean Bottéro's The Oldest Cuisine in the World, p. 28)
"Cut one pound of lamb shoulder into walnut-sized chunks or lamb stew meat and dice finely. Fill a medium pot halfway with water, and add the fat and the lamb. Add a teaspoon of salt; 12 ounces of beer; a finely chopped onion; a handful of arugula, finely chopped; ground coriander seed; and ground cumin. Bring the pot to a boil, and simmer for about one hour. Add in three peeled and quartered beets. Then, make a paste of one mortar. Add to the pot. Let simmer until the beets are tender, about 30 minutes longer. Sprinkle the soup with chopped fresh coriander before serving."

The following short video shows academics and their students recreating stews from the Yale collection with modern utensils.
https://www.youtube.com/watch?v=qfqhJNUtiww&t=30s

THE BABYLONIANS

The Babylonians were a post Sumerian culture which arose in southern Mesopotamia from about 1900 BCE, with the capital city of Babylon built on the Euphrates River. The timeline of this culture is divided into two distinct eras: the Old Babylonians starting from the time of Hammurabi (1792-1750 BCE) and the Neo Babylonian Empire of Nebuchanezzar in 587 BCE.

Hammurabi's code is a Babylonian legal text carved in cuneiform which was composed between 1755 and 1750 BCE. There were extensive laws dealing with the duties and debts of farmers (42-52), irrigation offences (53-56), cattle trespass (57-58) and the care of date orchards (60.)
Further laws dealt with agriculture such as oxen (241-252,) hire of agricultural implements (259-260,) hire of herdsmen (261) duties of shepherds (262-267) and the hire of wagons and beasts (268-272.)
REFERENCE: Wikipedia on Hammurabi's Code

Hammurabi's code in the Louvre, credit Mbzt, CC BY 3.0

In Babylonian times, farms occupied more land than in previous times when urbanisation was strong. Farmers lived in reed huts, isolated brick farmhouses or camps of tents. Some centres were fortified, while many areas had cisterns, granaries and threshing floors for the grain. Crop rotation was also practised to preserve the fertility of the soil. Trees were planted in rows to protect the farms from winds, and palm orchards were combined with smaller crops.

Salination presented a problem in Babylonian times due to the irrigation system which brought salt water to the crops. The Babylonians developed techniques such as the control of water through the canals, soil leaching to remove the salt and leaving fields fallow. Locust plagues were often a threat which had to be fought by destroying the larvae with water from the canals. They cultivated similar crops to the Sumerians, with sesame widely grown to make oil. Groves of date palms were planted, along with vegetables like onions, garlic and cucumber growing in the shade of these trees.

By the 18th century BCE, large parcels of land were privatised or leased from the temple. The smallest unit of land leased was the *ilkum* to small-holding families. As more land became privatised, landowners needed surveyors to establish boundaries and resolve disputes. The tape and rod were standard Babylonian surveying tools, and revered symbols of justice. Surveyor's plans have survived in clay tablets such as the 3,700 year old cadastral survey Si.427.

Recipe: From the Yale Babylonian Collection boasting three clay tablets in the Akkadian language that date to about 1750 BCE. A modern interpretation for a wild fowl pie contains this recipe:

- Ingredients: wild fowl, water, milk, salt, fat, cinnamon, mustard greens, shallots, semolina, leeks, garlic, flour, brine, roasted dill seeds, mint, wild tulip bulbs.
- Method: Salt fowl inside and out after de-feathering, and place in a pot with water and milk. When it comes to the boil add a mash of shallots, semolina, leeks, garlic and enough water to moisten. Cook until the meat is soft enough to debone, then remove and let cool. Roast dill seeds and remove from flame.

Andrei Mihai, "Archeologists reveal 5,000 year old Babylonian recipes." ZME Science, October 25, 2016.
https://www.zmescience.com/science/archaeology/babylonian-recipe-25112015/

THE ASSYRIANS

This northern Iron Age Mesopotamian Empire was founded on the Tigris River in the second millennium BCE. The Assyrians were more renowned for their military prowess than cultural achievements, but they learned to farm in a less hospitable environment than their predecessors in the south. They were able to cultivate fruits and vegetables in the cooler climate, such as almonds, firs, quinces and chestnuts. Barley was the principle cereal crop, along with minor crops such as emmer, millet, onions and flax. Their farming methods had advanced little since Sumerian times with oxen pulled ploughs, although mechanical seeders were developed. These ploughs had a small blade with a hopper which fed a chute that run down behind the blade, depositing seeds into the furrows.

Food and crops A scroll from the great library in Nineveh established by King Ashurbanipal (668-633 BCE) records many aromatic plants and herbs, including thyme, cardamom, saffron, garlic, cumin, poppy, coriander, dill and myrrh. Sesame was also utilised as a vegetable oil.

Another scroll describes the menu of a banquet for 69,574 people sponsored by Ashurbanipal when he opened a new palace. The ten day celebration included 5,000 foreign ambassadors as well as the king's subjects from all professions and classes. The menu shows that the banquet required 10,000 sheep, 34,000 types of fowl, 10,000 fish, 10,000 jars of beer and 10,000 containers of wine.

Although wine had been available since Sumerian times, most wine was probably imported from cities like Carchemish. Few texts inform us about wine production in Mesopotamia, although red and white wines were often listed.

Hydraulic engineering became more sophisticated as the Empire grew. Ancient Assyria was situated along the Tigris River which carried six times more silt than the Nile River. With shallow river beds which often silted up, the Tigris flowed faster than the Euphrates and often flooded dramatically, washing away the fields rather than replenishing them.

Canals and levees were built adjoining the river so that fields were irrigated by digging through the canal wall and shovelling mud into the breach when enough water had entered. The river flooded in spring and the canals were breached at this time on fallow fields to prepare them for planting. The fields would be ploughed in autumn after the rain, before the soil was rock hard, a process repeated for several passes.

Seeding commenced in October and was greatly improved by the mechanical seeder. The field was flooded again to leach salts from the furrows. In April barley and onions were harvested, followed by emmer wheat in June. Harvesting was done by hand with iron sickles, and the grain was threshed on a threshing floor by oxen dragging a heavy board over the grain.

One of the king's major duties was to maintain effective canal systems to water their lands. King Tiglath-Pileser I (1115 to 1077 BCE) boasted in an inscription:

> "I had plows put into operation throughout the whole land of Assyria, whereby I heaped up more piles of grain than my ancestors. I established herds of horses, cattle and donkeys from the booty which by the help of my Lord Ashur I had taken from the lands over which I had won dominion."
> Saggs, 'The Might that was Assyria" p 162

King Ashurnasirpal II, (883 to 859 BCE) who was famous for his military conquests also embarked on a major engineering project to expand the canal system around the nation's capital, Nimrud. He commissioned the digging of the Abundance Canal which linked Nimrud with the Upper Zab River. In another inscription he announced:

> "I dug out a canal from the Upper Zab, cutting through a mountain peak, and called it Abundance Canal. I watered the meadows of the Tigris and planted orchards with all kinds of fruit trees in the vicinity. I planted seeds and plants that I had found in the countries through which I had marched and in the highlands which I had crossed: pines of different kinds, cypresses and junipers of different kinds, almonds, dates, ebony, rosewood, olive, oak, tamarisk, walnut, terebinth and ash, fir, pomegranate, pear, quince, figs and grapevine." (ibid)

His canal also watered the royal gardens which were filled with many varieties of fruit.

King Sargon II (722 to 704 BCE) built a palace near the capital at Dur-Sharrukin (Korsabad) and planted an even bigger garden which required an expanded canal system. When his troops marched into the kingdom of Urartu in 714 BCE, they came across the Persian qanat system of using tunnels to bring the water from the hills to the plains. Sargon's qanawat system was built on this model and many are still in use.

Sargon's son Sennacherib (704-681) BCE moved the capital again to Nineveh on the Tebitu River. To supply the expanded city with water he dug a canal from the reservoir to the city. He also had an artificial swamp created to absorb water overflowing from the canals. This wetland was designated a royal game preserve which the king used for hunting deer, wild boars and water fowl.

Eventually eighteen canals were built to connect Nineveh to the Tebitu reservoir and other sources. Sennacherib's engineers dammed the Atrush River and included a self-operating sluice gate which allowed the water to automatically flow, a remarkable innovation. They had to build an aqueduct 30 feet high and 90 feet long of stone and sealed with concrete. The King boasted: "I caused a canal to be dug to the meadows of Nineveh. Over deep-cut ravines I spanned a bridge of white stone blocks. Those waters I caused to pass over it."
Lloyd, "Architectural Description of the Aqueduct," "Sennacherib's Aqueduct at Jerwan," 10-20, "The Ancient Engineers," 63-64.

Jerwan aqueduct, credit IbrahimKocher Duhok, CC BY-SA 4.0

King Sennacherib also rebuilt an old palace with massive terraced gardens which were watered by strange devices.

> "In order to draw water up all day I had ropes, bronze wires and bronze chains made, instead of shadufs I set up the great cylinders andalamittu-palms over cisterns…."

From inscription translated in Dalley and Olson, "Sennacherib, Archimedes, and the Water Screw: The Context of Invention in the Ancient World."

Historians today puzzle over these devices, with some historians like Stephanie Dalley concluding that the Assyrians invented a type of Archimedes screw which was not officially invented for another few centuries.

The Assyrians were some of the earliest people to build water fountains. An Assyrian fountain discovered in a gorge of the Comel River consists of basins cut in solid rock and descending in steps to the river. Small conduits led the water from one basin to the other.

Cuisine:

Gubibate is a modern Assyrian dish which is believed to go back to ancient times. It is made of bulgur wheat, onions, spices, ground beef, lamb or goat meat which can be baked, fried, cooked in broth or served raw. Today a similar recipe is called kibbeh.

Ashurbanipal at a garden party with grapes and palms, credit A Guck, CC BY 4.0

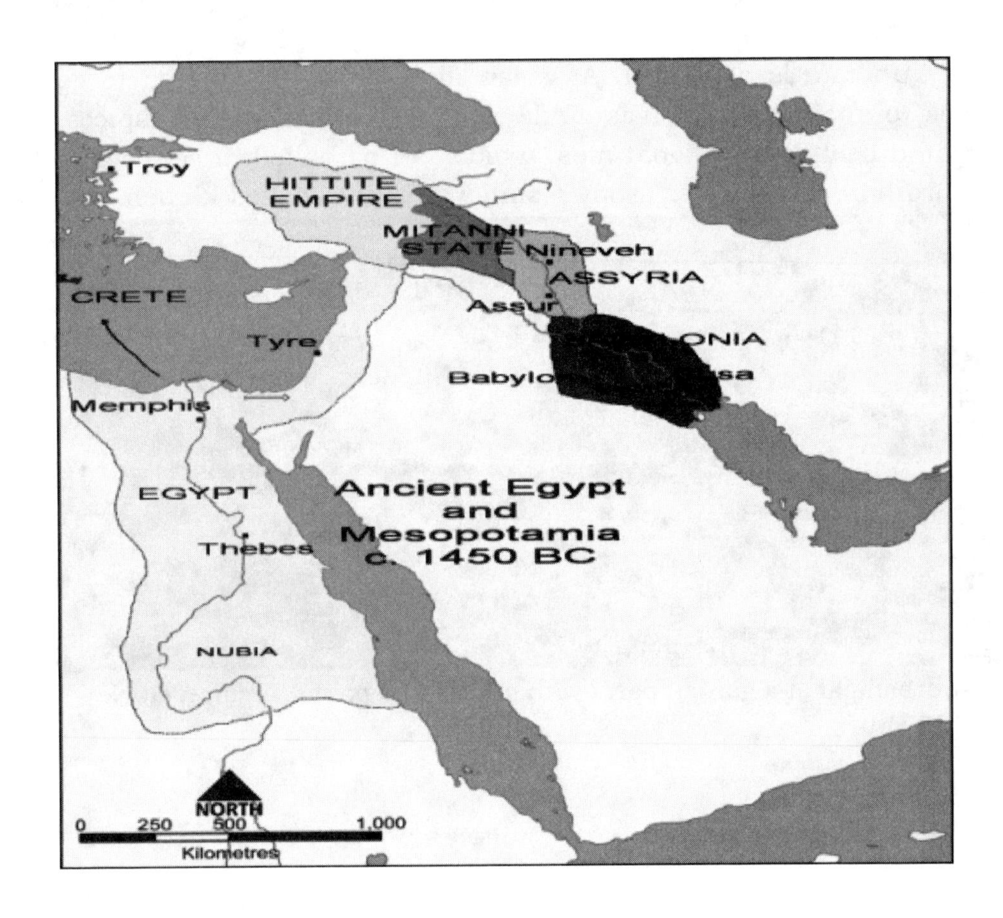

NEO BABYLONIAN EMPIRE

King Merodach-Baladan II of neighbouring Babylonia (721-710 BCE) was a fierce soldier and keen horticulturalist who wrote one of the earliest, treatises on vegetable gardens, with precise instructions on cultivating 64 plants including spices and herbs such as cardamom, coriander, garlic, thyme, saffron and turmeric.

The Hanging Gardens of Babylon, known as one of the seven wonders of the ancient world, were allegedly built by King Nebuchadnezzar II (605-562 BCE) for his Persian wife Amytis.

Although these legendary, terraced gardens do not appear in any Babylonian texts, they were described by Berossus of Kos, Strabo and Diodorus Siculus. The Hanging Gardens have also been attributed to the legendary Queen Semiramis who built a tunnel beneath the Euphrates River.

According to ancient texts, the gardens were built in tiers at the top of a citadel 60 stadia (3.6 km) in circumference, with brick walls 22 feet thick. The bottom tier was deep enough to plant the largest trees and watered by Archimedes screws from the Euphrates River. Despite the gardens being attributed by ancient scholars as one of the Seven Wonders of the World, no evidence has been found by archeologists.

Many historians now believe that the legendary gardens actually belonged to King Sennacherib or King Ashurbanipal of Assyria who both planted magnificent gardens at their palaces in Nineveh.

19th century etching of the Hanging Gardens of Babylon

IRAQ today is no longer the fertile centre of Mesopotamia, due to decades of conflict, displacement of farmers and corrupt management of resources. According to the Borgen Project,

> "Iraq has not focused as much on reforming its institutions governing agricultural industry networks. Iraq's State-Owned Enterprises are involved in every step of food production, processing and distribution. The government attempts to distribute food products and support the industry through its bloated Public Distribution System (PDS), which in 2019 cost $1.43 billion, and its yearly $1.25 billion effort to buy wheat and barley from Iraqi farmers at double the international price. Despite these expensive programs, Iraq still ends up importing 50% of its food supply.
>
> Inefficient growth, processing and distribution methods and a reliance on food imports place Iraq in a delicate position.

They are susceptible to global food chain supply network failures and the threat of a budget collapse due to the crash of oil prices. Such an occurrence would likely cause the food system to implode without the current level of government intervention. These governance issues, on top of decades of conflict and displacement, have exacerbated food insecurity in Iraq."

The pandemic of 2020-22 has worsened the levels of poverty and food insecurity. USAID has provided almost $240 million in emergency food assistance to Iraq since 2014, assisting with food baskets, cash for food and food vouchers under the coordination of the World Food Program, (WFP) a subsidiary of the UN Food and Agriculture Organisation (FAO.)

Bradbury, Connor, "Food Insecurity in Iraq," "The Borgen Project," August 18, 2020. https://borgenproject.org/food-insecurity-in-iraq/

THE PERSIAN EMPIRE

.

The ancient Persian Empire, also known as the Achaemenid Empire, was founded by Cyrus the Great in 550 BCE, and officially ended with the invasion of Alexander the Great in 330 BCE. It its height, it stretched from India in the east, to the Caucasus in the north and Greece in the west, making it the largest pre-Hellenistic Empire.

The early Persians were herders of sheep and goats because the land was poor for agriculture. However, the incorporation of Mesopotamia, the grasslands of Anatolia as well as Egypt and the Median Empire into their new empire made Persia an economic powerhouse. The invention of the qanat hydraulic system also allowed agriculture to flourish in arid areas.

Pasargadae was the great palace of Cyrus built in about 550 BCE with its magnificent gardens. The garden had a geometric plan with stone watercourses dividing it into quarters. Water channels defined the space between two palaces. It was planted with cypress, pomegranate, cherry as well as the lily and rose. The garden at Pasargadae is the earliest example of the *Chahar Bagh,* the originating principle of the Persian Garden which has persisted into modern times, with its four quadrants delineated by water courses.

The qanat system

The qanat irrigation system was developed in Iran during the Achaemenid Empire. The basic model was to build underground tunnels with evenly spaced wells in order to irrigate the parched fields. The first step was to identify a stream of water near the bottom of a mountain, after which a mother well, one metre in diameter was sunk. The muqni in charge of the well would dig at least 15 metres to the stream, but often much deeper. He would then determine the route and slope of the tunnel to be built, as the water had to flow at the correct speed. To calculate the slope a string and plumb line were used. The route was calculated using string and oil lamps which could mark where each well was to be built.

The tunnel was also built using the oil lamps as an underground guide to the next well. At the end of the qanat a pool or oasis was often formed to store water. The length of the qanats depended on the slope of the land and the distance of the village to the water source. In desert areas the qanats were longer. In the Kerman area 70 kms qanats were built over the centuries and 40 kms around Yazd. The Qanats of Ghasabeth are one of the oldest networks of qanats built during the Achaemenid Empire. The complex contains 427 water wells with a total length of 33,000 metres, or 20.5 miles.

The qanats had to be cleaned annually to avoid a build-up of discharge. During Achaemenid times, any man who built a new qanat or renovated an abandoned qanat had their tax waived for himself and five successive generations.

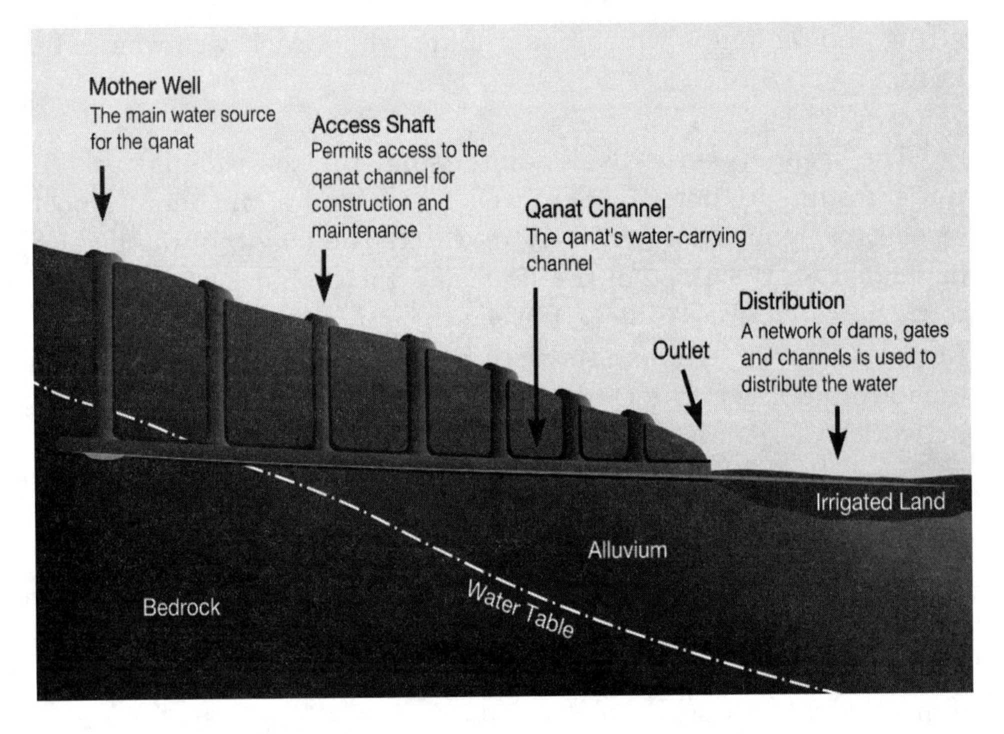

Qanat diagram, author Samuel Bailey, CC BY 3.0

FOOD The food grown and eaten by ancient Persians was varied and quite similar to modern Persian cuisine. Rice was an important crop introduced in the 5th century BCE, and was often combined with almonds, pistachios, lentils, saffron, onion, fennel, carrots and cilantro as well as meats such as lamb goat or chicken.

Bread was made from wheat or barley and could be combined with fruits or herbs. Different processes and ingredients made the bread leavened or unleavened, light or heavy. Fruits cultivated included melons, grapes, pomegranates, oranges and lemons. Fruit was often eaten as a dessert.

Persian wine was produced as early as 5000 BCE. The Persians were also fond of dairy. Cheese was popular as was yoghurt. A dried milk product called Kashk allowed for preservation of dairy ingredients for cooking.

Ice storage

By 400 BCE, the Persians advanced qanat technology to provide refrigeration and ice to the desert during summer. During the winter, ice was formed when an east-west wall channelled qanat water to the shaded north wall where it froze easily. From there it was stored in conical shaped yakchals which also acted as wind towers to draw cool subterranean air upwards from the qanat to keep the underground spaces cool even in the hottest summers.

Yakchal in Yazd Province, credit Pastaitaken, CC BY-SA 3.0

The Persian Empire included economically developed countries such as Mesopotamia, Egypt, Syria and Asia Minor as well as diverse tribes such as the Libyans and Lycians. Crops grown across the empire included:

- Babylonia- wheat, barley, spelt, millet, sesame, peas, mustard, garlic, onions cucumbers, apricots, apples, pomegranates, dates, wine, vinegar and honey, beer brewed from barley, livestock
- Elam- barley, wheat, wine, vinegar, beer
- Egypt –barley, livestock
- Palestine (Levant) wheat, peas, lentils and mustard

Land ownership across the empire was of several kinds, from vast estates in Babylonia leased out in large plots to major tenants who sublet them in smaller plots, to lands confiscated from conquered rulers, particularly those countries which had not willingly subjugated themselves to the Achaemenids. The Persian king also owned many irrigation canals, forests in Syria, fishing rights from Lake Moeris in Egypt as well as gardens and palaces across the empire.

Persian recipe

No recipes from the Achaemenid era survive, but recipes from the Parthian period from 274 BCE to 224 CE were preserved by Roman sources principally *"De Re Coquinaria"* by Marcus Gavius Apicus. "*De Agricultura*" by Marcus Portius Cato had recipes for Parthian Bread, Parthian Chicken and Parthian Lamb.

Parthian Lamb recipe adapted:

- ½ small milk fed lamb or kid
- 1.5 pounds of pitted prunes plumped in water or wine
- 4 large chopped onions
- 2 teaspoons garum (fish sauce)
- 4 tablespoons olive oil
- 3 garlic cloves
- Glass of white wine
- Pepper

Directions

- Score lamb and half chop through bones
- Rub lamb in olive oil, garlic, salt and pepper
- Roast lamb slowly at 170 degrees C or 335 degrees F for 1.5 to 2 hours, basting slowly with the wine.
- Sauté onions, add salt, pepper and herbs. Then add prunes and pureed garlic, and finally add garum. Remove from heat
- After removing lamb, sprinkle with a little white wine to degrease.
- Place in oven for 10 more minutes and serve with pepper.

http://cultureofiran.com/ancient_iranian_recipes.com

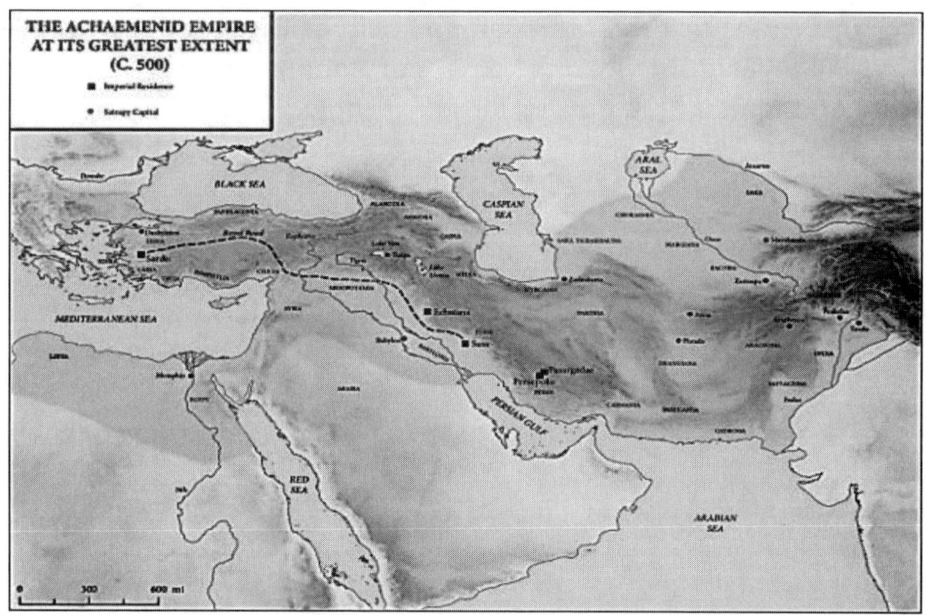

Persian Empire, credit Mossmaps CC BY-SA 4

IRAN today is facing food insecurity on multiple fronts. In May 2018, the ILNA newsagency quoted a parliamentarian warning that food security in the provinces of Sistan, Baluchestan, Hormozgan, Bushehr and Southern Khorasan was on the verge of a crisis.

ISNA, another state run newspaper reported that some residents in these provinces were reduced to eating cat's meat, and that malnutrition was widespread.

In 2022, the price of staples increased dramatically due to supply chain problems resulting from the pandemic. Foods such as chicken, eggs, milk and cooking oil increased in price by up to 300%, causing widespread protests across the country.

According to blogger Alam (pseudonym) living in Khuzestan,

"A week ago, a half kilo of pasta was around 12,000 toman [0.4 euros], but now it's 28,000 [0.93 euros]. Oil was 120,000 [4 euros] for four litres and now is 400,000 [13.3 euros]. If you can find it, bread is twice as expensive as before. These are – or were – our staple food ingredients. And our family revenue is about 6,000,000 toman [200 euros], for four people."

Ershad, Alijani, "They are imposing famine on us. Soaring food prices fuel angry protests in Iran," "The Observers," May 13. 2022.

https://observers.france24.com/en/middle-east/20220513-they-re-imposing-famine-on-us-protests-spread-among-iranians-faced-with-soaring-food-prices

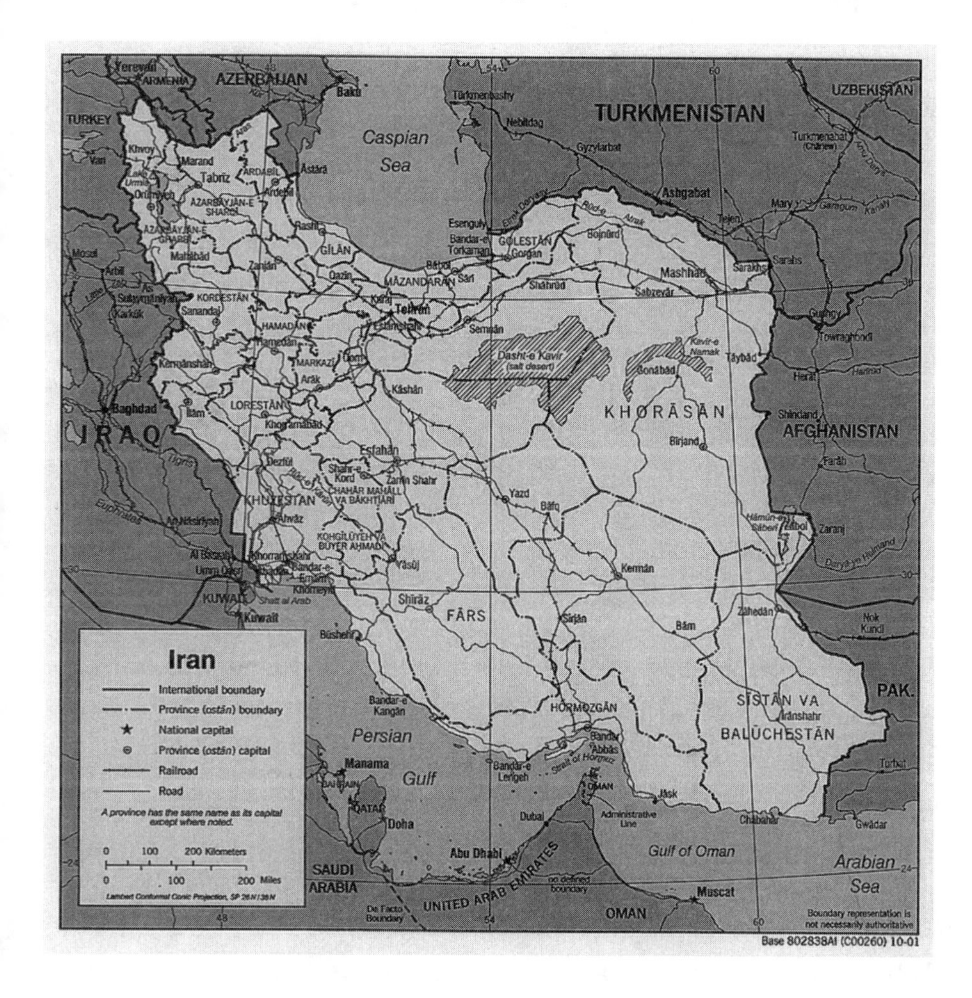

THE HITTITES

The Hittites were an Anatolian culture which ruled central Turkey from the 18th century BCE to 12th century BCE. In their heyday, the Hittite Empire challenged both Egypt and Babylonia for leadership in the region.

The landscape is covered by a highland plateau with deep river valleys surrounded by mountain ranges, which meant the temperatures were extreme. Anatolia lacked the great river valleys of Mesopotamia or Egypt, but the smaller river valleys were well-watered and fertile. The Hittites grew grains, herded animals and engaged in widespread trade.

On October 20, 2020 "Sci News" announced the recent excavation of a royal storage complex at the Hittite capital city of Hattusa. The site was first excavated in the 1930s by archeologists from the Deutsches Archäologisches Institut who later unearthed a massive subterranean storage complex measuring 118 m in length and 33-40 m in width in 1999.

The capacity, between 7,000 and 9,000 m3, was large enough to store up to 7,087 tons of cereal grain, enough to feed a population of 20,000 to 30,000 for one year, according to Professor Amy Boggard of The University of Oxford. The silo was divided into 32 individual storage chambers that were hermetically sealed and could be emptied independently. Some of the silo's chambers were full, containing hundreds of tons of intact charred cereal grain. The charring probably occurred during a fire which took place in the early 16th century BCE.

The contents were varied, containing well-prepared cereal grains, cereal chaff, pulse and weed seeds. Grains consisted of hulled barley, spikelets of hulled wheat, emmer and einkorn wheat. Many species of weeds were included, such as wild bishop and corn buttercup which are native to the Anatolian plateau. Other plant species identified from the silo include free-threshing wheat, bitter vetch, lentil, grass pea and a variety of small-seeded broad bean.

According to the article which quoted a paper reviewed in "Antiquity 94," "Hittite farmers fulfilled their tax obligations by engaging in low-input production of cereals that provided reasonable yields even under marginal growing conditions."

They concluded, "The detailed reconstruction of Hittite agroecology suggests that large-scale, extensive cereal production was a key state-sponsored economic strategy, with implications for the promotion of land-based wealth inequality and the territorial expansionism of many ancient states."

Enrico de Lazaro, "Ancient Hittite Farmers Paid Taxes in Barley and Wheat," "Sci News," October 20, 2020.

http://www.sci-news.com/archaeology/hattusha-silo-08966.html

Food and crops The Hittites enjoyed a nutritious and varied diet.

- Cereals were the staple crop, with barley and emmer wheat the most popular grain for baking bread in ovens, or on baking plates. Various types of bread were made with and without yeast. Pastries contained fruits, oils or honey.
- Fruits and vegetables: apple, date, fig, grape, olive and pomegranate, vetch, broad beans, chickpeas, lentils, cucumber, pea, garlic and leek.
- Fats used for frying foods were mainly from animal sources, as well as olive oil, flax, sesame oil, nut oil and black cumin.
- Dairy products from cows and goats included milk, cream, butter and sweet milk.
- Alcoholic beverages: beer made of barley or wheat beer-bread called bappir and fermented in large vats. Wine was drunk at the Hittite court and often used as libations to the gods.
- Animals were bred for food, milk, clothing and sacrifice rituals. They included cattle, sheep, goats, pigs.

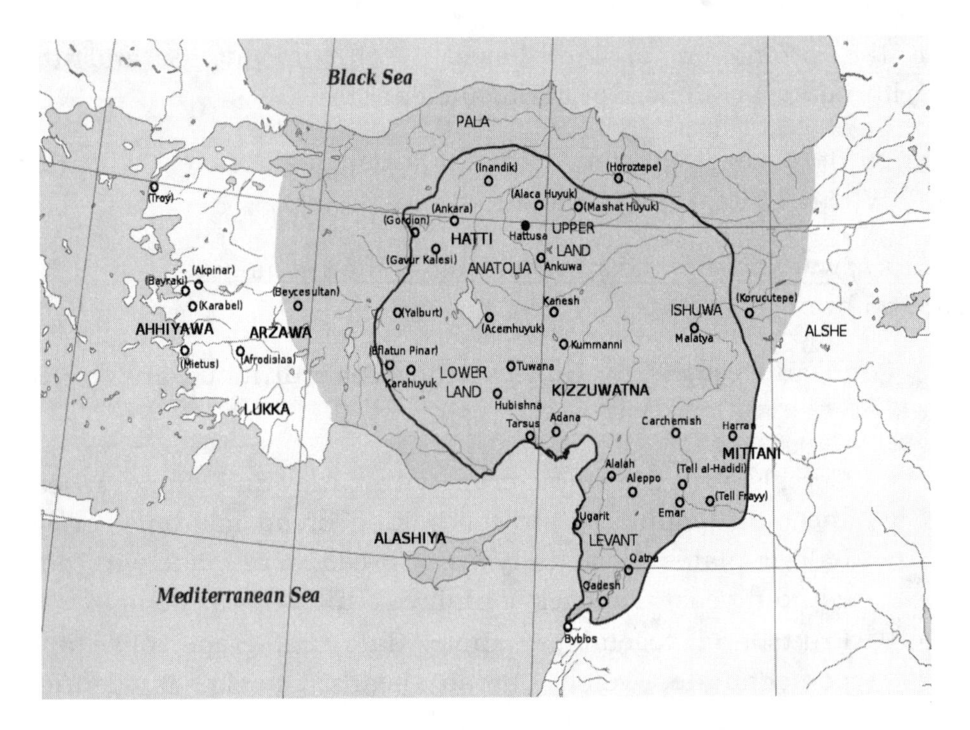

Hittite Empire at its height. CC BY-SA 3.0

CUISINE

The Hittites left numerous texts detailing food preparation, using utensils such as storage pithoi, cooking pots, tripod pots, stewpots, kettles, braziers, cookers, baking plates, hearths, ovens and vessels for milk and cheese processing.

Several laws were concerned with sanitary conditions in kitchens and penalties were particularly harsh. “If a chef with a large, unmanaged beard or long, unmanaged hair cooks in the kitchen or an animal wandered into the kitchen, he or she used to receive a death penalty along with their family,” according to the “Daily Sabah.”

Aykut Çınaroglu, the head of the excavations at Alacahoyuk, and professor of archaeology at Ankara University, told Anadolu Agency (AA) that Chef Omur Akkor, an excavation team member, prepared a special Hittite menu based upon ancient Hittite tablets in 2015. The chef used methods and ingredients available to Hittites and made yeast free bread using pounded wheat not sifted flour, cold meat and cooked onion.

Using German buckwheat which was crushed on stones, the only kitchenware used was a knife. Akkor claimed that more than 100 pastries were found on Hittite tablets, as well as olive oil, honey, beverages and vegetables. He cooked numerous breads, apricot butter, beruwa (mashed food) with cucumber, beruwa with chickpea, happena (a meat casserole with olive oil and honey,) kairya (grilled lamb liver and heart, cold meat, and meat with onion sandwiches.

"Archeological team prepares 4,000-year-old Hittite meals," Daily Sabah, September 9, 2015.

https://www.dailysabah.com/food/2015/09/09/archaeological-team-prepares-4000-year-old-hittite-meals

Hittite feast, credit "Daily Sabah"

Across the Mediterranean from Italy to Anatolia and the Levant, a steep decline in annual rainfall occurred during the 13th and 12th centuries BCE.

The downfall of the Hittite empire coincided with, or was caused by a devastating three year drought from 1198-1196 BCE. Crop failures would have led to a collapse of the tax base and general discontent. Soon after the capital Hattusa was burned and abandoned.

THE LEVANT- SYRIA, ISRAEL AND LEBANON

This strategic area of the Middle East, the western arm of the Fertile Crescent, was one of the earliest agricultural homelands during the Neolithic era. It also produced many cultures from the Bronze Age, such as the Canaanites to the Phoenicians and Hebrews.

THE PHOENICIANS were a maritime race centred in Syria and Lebanon from 1500 to 300 BCE. Because the land was not very arable, the Phoenicians set up extensive trading links and colonies around the Mediterranean, including Carthage in modern Tunisia. Phoenicia was famous for its cedar forests and purple dye obtained from the murex snail.

Food Unfortunately, no written records from the Phoenicians describing agriculture or food have survived, but their diet was similar to that in Mesopotamian countries with added imports. Grains such as wheat and barley were very important and often were imported from Egypt. The Phoenicians baked bread and flatcakes. Breakfast may have been a type of porridge made from cereals supplemented with cheese, eggs and honey called puls.

They also ate stews made with meat as well as lentils, chickpeas and beans. Meat would include sheep, rabbit, cattle, chicken, game and doves which could be roasted, grilled or made into a stew. Fish and seafood were abundant. Fruits included figs, pomegranates, dates, apples and olives. Herbs and spices were plentiful with salt the main condiment. Phoenician wine grown from local grapes turned into raisins was also highly regarded in ancient times.

Carthage became a large and prosperous colony which threatened Rome, leading to the Punic Wars. After the third Punic War in 133 BCE, the Romans razed Carthage to the ground and sold its citizens into slavery.

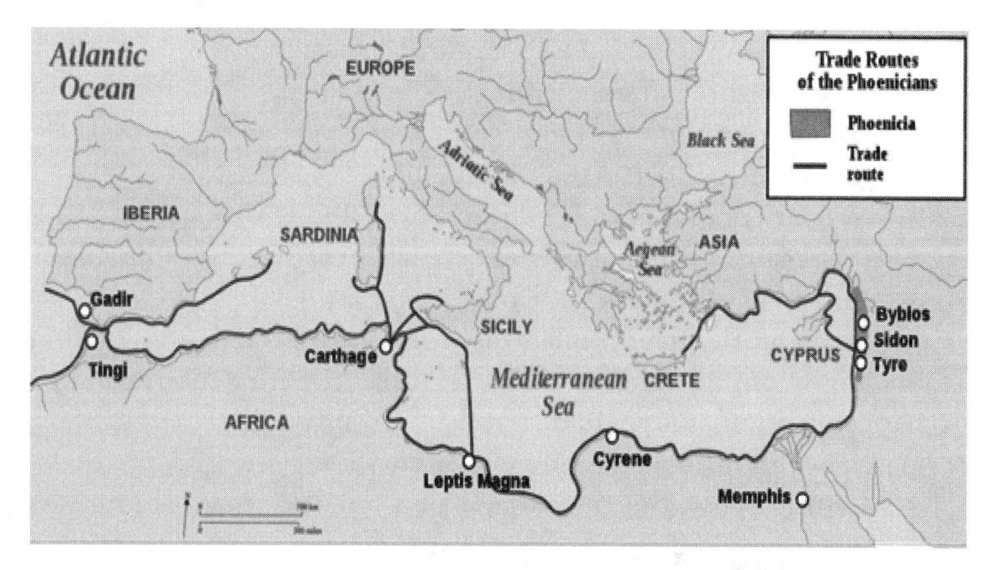

Phoenician trade networks 1200-800 BCE. CC BY-SA 3.0

Agriculture Carthage in North Africa was considered the granary of the central Mediterranean. The colony became very prosperous as it expanded Berber knowledge of agriculture and animal husbandry with Phoenician innovations. Its trading partners extended into Britain for tin, Spain, Sicily and Cyprus.

Carthage was surrounded by two rings of agriculture; the inner ring was used for olives, fruit trees, grapes and vegetables. Farm animals and horses were bred there, as were bees. Oxen were highly regarded draft animals for pulling iron ploughs, providing manure for the fields and oxhides.

Mago the Carthaginian was referred to as the Father of Agriculture by the Romans who destroyed his city and ploughed salt into the fields to curse the land in 133 BCE. The Romans rescued his 28 page treatise and fortunately translated it into Latin. Today, fragments survive in quotations by Roman writers on agriculture, including Varro, Pliny the Elder and Columella. According to Wikipedia these fragments survive.

- The most productive vineyards face north- Columella, *De agicultura*
- How to plant vines.-Pliny, *Naturalis Historia*
- How to prune vines- Columella, ibid
- How to plant olives- Pliny, ibid
- How to plant fruit trees- Pliny, ibid
- How to harvest marsh plants- Pliny, ibid
- Preparing various grains and pulses for grinding- Pliny, ibid
- How to select bullocks- Columella, ibid
- Notes on the health of cattle- Varro *De Re Rustica*
- Mules and mares foal in the twelfth month after conception- Varro, ibid
- Notes on farmyard animals- Varro, ibid
- Getting bees from the carcass of a bullock or ox- Columella, ibid
- The beekeeper should not kill drones- Columella, ibid
- How to preserve pomegranate- Columella, ibid
- How to make the best *passum* (raisin wine) Columella, ibid

This recipe for turning wheat and barley into flour is attributed to Mago.

> "Soak the wheat in plenty of water and then pound it with a pestle dry it in the sun and put it back under the pestle. The procedure for barley is the same. For 20 parts of barley, you need two parts of water."

Carthaginian raisin wine was sold to Italy in huge quantities.

> "Pick some well-ripened early grapes. Discard any that are mildewed or damaged. Drive forked branches or stakes made of rods tied into bundles into the ground at a distance of about 4 feet apart. Lay reeds across them and spread the grapes out in the sun on top. Cover them at night so that the dew will not moisten them. When they are dried, pick the

grapes off stems and put them in a jar or pitcher. Add some unfermented wine, the best you have, until the grapes are just covered. After six days, when the grapes have absorbed it all and are swollen, put them in a basket, put them through the press, and collect the resulting liquid. Next, press the mass, adding fresh unfermented wine made with other grapes which have been left in the sun for three days. Stir it well and put it through the press. Bottle the liquid of the second press in stoppered jars, so it will not turn sour. After 20 or 30 days when the fermentation is over, decant it into fresh vessels. Coat the lids with plaster and cover them with leather."

Source; "Mago of Carthage is the father of Agriculture and Farming." https://phoenicia.org/Mago-Carthage-Father-of-Agriculture.html

SYRIA has gone from being relatively self-sufficient in food security a decade ago to featuring among the top ten worst food crisis ridden nations in 2022. After achieving food sufficiency in 1994, agriculture was one of the country's main economic sectors, and Syria enjoyed the most thriving agricultural sector in the Middle East.

War has devastated the country since 2011 along with starvation, water insecurity, sanctions and electricity shortages, but the Ukraine war and rampant inflation have made the situation much worse.

In 2022, acute food insecurity is affecting 12 million people, an increase of 51% since 2019. Of the 113 countries assessed on the Global Food Security Index, Syria is ranked at 106. Sanctions imposed upon Syria still allow for the importation of food, but the banking system and shipping sector have made the process difficult. Prohibitions on importing fertiliser, pesticides and herbicides as well as agricultural equipment caused prices to skyrocket.

According to Mohammed Kanfash of the World Peace Foundation,

> "Sanctions contribute to the persistent and ever-deteriorating food insecurity crisis in Syria. Although they are not to blame solely or mainly for the crisis, they do play a significant role and hence their use should be revisited and reviewed in light of Syrian food security concerns, as recommended by scholars and researchers on topic."

Kanfash, Mohammed, "Sanctions and Food Insecurity in Syria," World Peace Foundation, July 6, 2022.

https://sites.tufts.edu/reinventingpeace/2022/07/06/sanctions-and-food-insecurity-in-syria/

LEBANON'S food insecurity has suffered since the huge influx of refugees from Syria in 2015. Security for Syrian refugee households has fallen from 25% to 11% in 2020, making them dependent upon assistance provided from UNHCR.

Lebanon imports over 50% of its food, with the majority of sunflower oil and wheat originating in Ukraine and Russia prior to the Ukraine war of 2022. Local agriculture suffers from mismanagement of water resources, antiquated equipment and corruption.

In 2020, a devastating explosion destroyed the wheat silos which held the strategic stockpile of grain for Beirut. This stockpile held enough grain to last for three months and was concentrated in one place. The destruction of the silos forced Lebanon to rely on wheat imports of 30,000 to 35,000 tons per month to produce the flour to make bread. The Lebanese Minister of Economy, Amin Salam, urged the Emir of Kuwait to rebuild the silos in 2023.

In September 2022, Lebanese banks shut indefinitely after a spate of bank heists, often by depositors trying to get access to their own savings. The economic meltdown which began in late

2019 has seen the country's gross domestic product contract by more than 40% since 2018, and the currency has collapsed.

By 2023, around two million Lebanese and Syrian refugees face food insecurity, nearly one third of the total population.

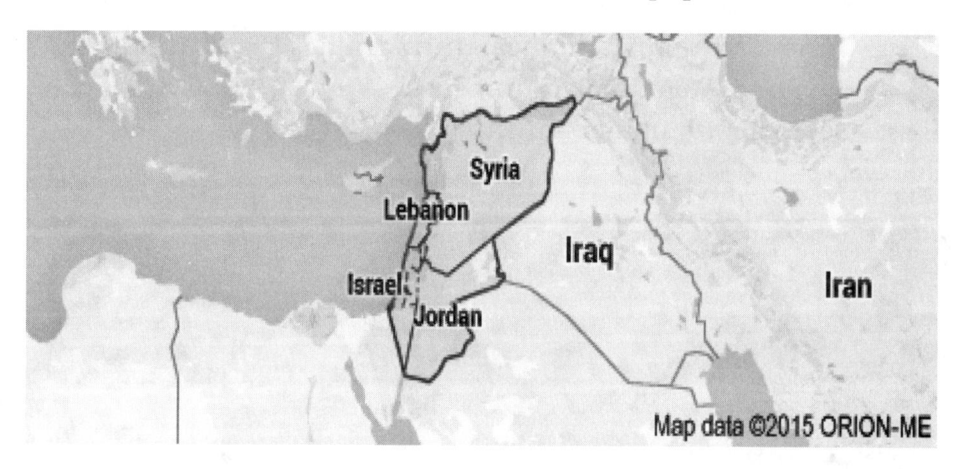

ISRAEL/PALESTINE

Wild barley and emmer wheat were domesticated and cultivated in the Jordan River Valley from about the 9th millennium BCE, as well as einkorn, emmer and barley in the Neolithic city of Jericho on the West Bank. The development of pottery from 6000-4300 BCE enabled people to store and transport both food and liquid. An agrarian and herding economy developed, with crops such as figs, lentils and broad beans found in the archeological record. From 4300-3300 BCE large pots were fired and the earliest date palms were cultivated at Ein Gedi by the Dead Sea. Olives were grown in the Golan and an olive oil industry developed.

During the Bronze Age from 3300-1200 BCE wine and olive oil became important exports to Egypt and Phoenicia from the Canaanites. The Gezer calendar written in Canaanite, describes monthly periods to harvesting, planting and tending specific crops. According to the text:

- October and November for gathering.
- December and January for planting.
- February and March for late sowing.
- April for cutting flax.
- May for reaping barley.
- June for reaping and measuring grain.
- July and August for pruning.
- September summer fruit.

Gezer calendar. CC BY_SA 3.0

Stone lined storage pits for grain have been discovered from this period. Around 1200 BCE, the Canaanites began to excavate sophisticated tunnels for water which was used for domestic as well as irrigation purposes. A Canaanite tunnel near the later city of Jerusalem collected the spring water as early as c1800 BCE. It extended southwards through the bedrock, and then released the water onto the fields in the Kidron Valley before ending up in an open reservoir. This irrigation expanded the arable land in late Canaanite times.

THE ISRAELITES, also known as Hebrews, occupied the earlier land of Canaan from about 1200 BCE, during the early Iron Age, until Roman times. Although their diet was similar to that of others in the Levant, they were the first known culture to institute dietary restrictions based upon religious beliefs. Literary sources include the Hebrew Bible Old Testament, the New Testament, the Dead Sea Scrolls, the Mishnah and Talmud. These dietary laws are still practised today by many of the Jewish faith and include kosher requirements which are based upon the Book of Leviticus.

Dietary staples included bread, wine, olive oil, honey, legumes, fruits, vegetables, dairy products, fish and meat. While pastoralism and animal husbandry remained important, the construction of terraces in the hills, additional plastered cisterns for water storage and tunnel building such as Hezekiah's tunnel in Jerusalem, increased the food supply by providing water for irrigation.

Farming tools included the ard, the hoe and the mattock, a heavier tool for breaking up the soil. The iron mattock could penetrate the fields for a few inches and was pulled by draft animals such as oxen, horses or donkeys.

In Hellenistic times, (3rd century BCE) the author of the Letter of Aristeas (pars. 112–118) praised the agricultural productivity of the country and the great "diligence of its farmers. The country is plentifully wooded with numerous olive trees and rich in cereals and vegetables and also in vines and honey. Date palms and other fruit trees are beyond reckoning among them." However, this author may have exaggerated the extent of the irrigated areas, the importance of the Jordan River and the size of the landholdings.

DIETARY RESTRICTIONS of KASHRUT

Before the time of Moses the only dietary restriction was against the consumption of blood. The Hebrews believed that blood contained the 'life' of the creature and that all life belongs to God. This blood prohibition is still in effect among Jews so that animals have to be slaughtered and the blood drained to be kosher.

Jewish Dietary Laws were outlined in Leviticus 11 and Deuteronomy 14 which set the ancient Israelites apart as a separate people. They were believed to have been handed down by YHWH to Moses. Both accounts share the basis of the food restrictions.

- For animals that live upon the ground, the Israelites are not to eat those animals, which do not both split the hoof or chew the cud (Lev. 11:3; Dt. 14:6). Leviticus lists four animals, which are forbidden within this category; the camel, rock badger, hare, and the pig (11:4-8). Deuteronomy shares the same list, however it includes animals, which may be eaten; the ox, sheep, goat, deer, gazelle, roebuck, wild goat, ibex, antelope, and mountain sheep. (14:4-5)

Meat was usually boiled, roasted or cooked in a stew. At no time could a kid be boiled in its mother's milk (Deut: 14:21) which was apparently a Canaanite practice. This regulation later became a prohibition against even eating meat and milk products together under kosher law.

- As for sea creatures, only those which have fins and scales may be eaten, according to Leviticus 11:9-12. This prohibits shellfish and creatures like squid or octopus.
- Land creatures which crawl or creep such as lizards and snakes are forbidden.
- 20 categories of bird are forbidden, such as birds of prey, but Deuteronomy adds that the Israelites can eat a clean

bird, which is undefined. (14:11-20) These fowls were probably chicken, duck and goose whose eggs could also be eaten.

- Insects such as the locust, cricket and grasshopper are acceptable.

The second part of the Food laws concerns the prohibition of contact with unclean foods. These prohibitions include touching a clean or unclean animal, including a dead creeping animal.

Whilst the Israelites were not the only people of the area to prohibit pork, they are the only people with surviving texts which prohibit its consumption. The reason for prohibiting pigs is speculative, but may result from the fact that swine are scavengers which are not easy to herd.

The Old Testament had many other references to food and agriculture. Burnt savoury meat was offered up to YHWH as a sacrifice. The Passover is a food ritual from the time of Moses with unleavened bread prohibited and only matzo allowed. Hannukah which dates from the Greek occupation, has its own food rituals, particularly frying in oil.

Passover Seder is a ritualised meal originating in the Book of Exodus. Matzo, unleavened bread replaces yeasty bread for 7 or 8 days. A seder plate contains ritually prescribed foods which commemorates the story of being freed from bondage in Egypt. Maror are bitter herbs like horseradish, charoset is a thick paste of fruits and nuts to recall the mortar used by slaves, karpas is usually parsley dipped into salt water to commemorate tears, z'roa is a roasted lamb bone commemorating ritual sacrifice in the temple and bietzah is a roasted egg a symbol of mourning. Gefilte fish is also a common Jewish recipe.

REFEENCE: Albala, Ken "Food: A Cultural Culinary History" "The Great Courses," 2013.

Recent archeological evidence, however, indicates that ancient Judeans were eating non-kosher fish like catfish throughout the first millennium BCE in defiance of the Mosaic laws of kashrut. By Roman times (63 BCE-324CE) such fish disappeared from the archeological record. Moreover, sites from the ancient Kingdom of Israel in the south also yielded pig bones dating to the 8th century BCE. It is very possible that kashrut laws were only enforced during the Roman occupation.

RECIPE: Ashishim, a red lentil pancake, was a common dish eaten by Jews in antiquity. It is made of red lentils, eggs, flour and sesame seeds which are mixed together to create a batter which is deep fried and topped with honey syrup.

This recipe for a similar dish, ashishot, also extends back to ancient times.

Ingredients: 8 oz red lentils

1 tbs whole wheat flour

3–4 tbs honey

½ tsp cinnamon

¼ cup olive oil

Method: 1 Toast lentils in fry pan and grind into flour.

2. Mix lentil flour with olive oil making small balls of dough which are flattened into small pancakes.

3. Using a frying pan, fry pancakes gently on both sides until golden.

REFERENCE: Vamos, Miriam Feinberg, “Ancient Hebrew Sweet you Could Make Today, Ashishot,” “Haaretz,” April 23, 2021, tinyurl.com/w2jx64st

Modern seder plate

THE ARAB ISLAMIC EMPIRE 750-1258 CE saw the rapid conquest of the Middle East, North Africa, Spain and Persia by Arabs from the Arabian Peninsula, and the institution of Islam as the state religion. The cuisines of many ancient Mediterranean cultures were adopted. During the rule of Abbasid Caliph Harun al-Rashid from 786-809, the capital was moved from Damascus to Baghdad where the golden age of Islamic science flourished with the rediscovery of ancient Greek writings, and the wealth generated by trade from China, Africa, Arabia and India.

Food *Harisa* was not the spice used today but a kind of porridge made from pounded wheat, butter, meat and spices. It became popular during the 7th century in Damascus and has been preserved by Yemeni Jews to this day. Another popular medieval Arab dish is *tharida*, a crumbled bread soup moistened with broth. It was said to be one of the Prophet Muhammed's favourite dishes originating from the Qurayish tribe of the

Arabian Peninsula in early Islamic times. *Asida* is a semolina porridge, rooted in the culinary traditions of Muslim Andalusia. Later, it became a broth with wheat flour added as well as honey and butter. It was fed to women in labour and eaten on religious holidays. *Rafis* is another dish made of wheat flour, dates, honey, butter and spices.

Cookbooks were in great demand during Abbasid Baghdad. **Ibrahim al-Mahdi** was one of the most passionate culinary writers, but unfortunately only fragments of his cookbook survived in other texts. Al-Warraq's cookbook "Book of Dishes" was a full-fledged volume with 132 chapters comprising in excess of 600 recipes. Little is known about **al-Warraq**, who was commissioned to write a cookbook on the cuisine of Caliphs, lords and dignitaries, including much older recipes. His cookbook contained five chapters about kitchen utensils, spices, eight types of tastes, remedies for burnt food, why food spoiled; 70 chapters on recipes, 20 chapters on cooking and dining etiquette and 25 chapters on medicinal properties of food.

His hangover cure, "*Kkushkiyya*" is a meat, chickpea and vegetable stew with a special ingredient known as khask, a fermented yoghurt milk product. Al-Warraq also advised eating cabbage prior to imbibing and eating snacks between drinks to slow down alcoholic effects.

At the same time, new eastern crops such as sugarcane, rice, taro, eggplant and citrus fruits were introduced. Lemons were brought to Syria, Palestine and North Africa. Ibn al-Awwam's "*Kitab al-filaha*," (Book of Agriculture) devotes an entire chapter to the lemon. The texts reveal that the Arabs improved irrigation systems using canals and cisterns while introducing new cultivation techniques, which allowed for improvement in harvesting existing species and the introduction of new ones.

The Islamic conquest of al-Andalus in 711 CE brought many new crops to the Spanish peninsula such as rice, eggplants, saffron and durum wheat which became the principal source of semolina wheat for bread, pasta and couscous. The following century, figs replaced the Arabic dates and all dishes were cooked with olive oil, rather than sesame oil.

Agriculture in the Arab empire was built upon previous methods inherited from the Romans, Hellenistic Greeks and Persians. Agronomy books by Ibn Bassal and Ab□ l-Khayr al-Ishb□l□, demonstrate the extensive diffusion of useful plants to Medieval Spain (al-Andalus.) The twelfth century agronomist Ab□ l-Khayr al-Ishb□l□ of Seville described in detail in his *Kit□b al-Fil□ha* (Treatise on Agriculture) how olive trees should be grown, grafted treated for disease, and harvested. In addition to describing his own experiments, he also gave similar detail for crops such as cotton.

Horticulturalists such as Ibn Bassal wrote detailed treatises on how to propagate the olive and date palm, crop rotation of flax with wheat and barley, and companion planting grapes with olives. Irrigation expanded due to the growing use of water and wind power. Windpumps were used to pump water since the 9th century in Persia and Afghanistan. Animal powered irrigation using the sakia wheel was introduced to Islamic Spain in the 8th century. In the Fayyum depression of Egypt, large scale irrigation projects with gravity-fed canals were undertaken.

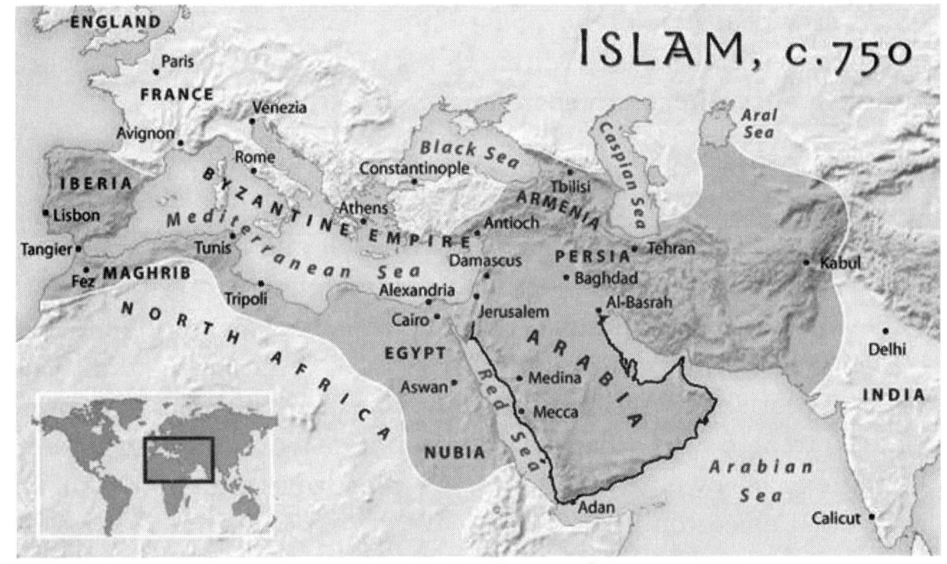

Arab Islamic Empire by 8th century CE.

Barida is a medieval Arabic recipe with ancient Roman roots. It includes murri, a fermented barley sauce very similar to Roman garum. Al-Warraq's "Book of Dishes" features a number of recipes for barida, fish or chicken topped with a complex sauce of herbs and spices. Although Al-Warraq traced it to the 8th century Abbasid caliphate, its roots extend to Roman North Africa.

Modern food author Lilia Zaouali provides this recipe for barida in her 2009 book "Medieval Cuisine of the Islamic World."

RECIPE: Cold Chicken with Spices and Herbs

1. Season chicken legs with salt and grill over medium heat

2. Take some vinegar and murr□ and in them macerated coriander [seeds], Chinese cinnamon, pepper, dried and fresh thyme, cumin, caraway, fresh coriander [cilantro], mint, rue, celery, the pulp of a cucumber, and elecampane. Put everything in a grinder, mix, and pour over the grilled chicken.

Serve cold.

3. Elecampane is the edible root of sunflower. Rue is a bitter herb which was popular in ancient times. Dandelion leaves can be a substitute for both.

Al-Warriq's book referred to three different types of carrots which were used for aphrodisiacs and diuretics; red, yellow and white carrot. One aphrodisiac recipe:

> "Slice carrots as thin as coins, put them in a pot with an equal amount of water. When cooked, strain the liquid and discard the dregs. In a clean pot, combine two parts of strained carrot juice and one part honey. Boil the mixture until one part evaporates. Add a small amount of mace and nutmeg. Set aside for days and use it. It is a beneficial drink."

REFERENCE: "Annals of the Caliphis kitchens English translation of Ibn Sayyar al-Warraq's tenth-century Baghdadi Cookbook."

PART 3

AFRICA & EGYPT

EGYPT

WEST AFRICA

SOUTHERN AFRICA

ETHIOPIA

CENTRAL AFRICA

EGYPT had extensive bread and beer production before the 1st Dynasty.

FERMENTED BREAD with a culture began in Egypt.

SOUTHERN AFRICAN kingdoms were often cattle economies.

CROPS from Africa- sorghum, types of millet, watermelon, black eyed peas, okra.

COFFEE was first grown in the Ethiopian Highlands.

EGYPT

One of the earliest agricultural civilisations, the Egyptians, were entirely dependent upon the Nile River annual flooding which deposited fertile soil in a process known as the **inundation.** In order to take advantage of the river, the Egyptians invented **basin irrigation** which involved building canals, ditches and earthen walls. The earliest reference to irrigation in Egypt was found on the mace head of the Pre-dynastic Scorpion King who ruled about 3100 BCE. It showed the king cutting a ditch for irrigation.

Basin irrigation allowed the Egyptians to control the rise and fall of the river using earthen walls to trap the water in one basin until it was drained to another basin in need of water. A network of earthen walls was built by farmers on their own land, as there didn't appear to be many state sponsored irrigation projects.

The Nile would regularly flood in August and September, leaving the floodplain and delta submerged by 1.5 metres of water. When the waters receded in October, farms were left with well-watered and fertile soil to plant their crops. This was called Peret. These crops ripened and were harvested in Shemu, between March and May. The Egyptians grew many crops for consumption, including grains, vegetables and fruits.

The personification of the inundation was a minor god called Hapi who made offerings of water and other foods to the pharaohs. Osiris was also associated with the Nile and fertility of the land. Other deities included Nebri, god of grain depicted with grains drawn on his body; Sekmet, goddess of the fields who often appeared with offerings such as eggs and ducks and Hathor, the cow goddess.

Tools The ox drawn plough was designed in two gauges; light and heavy. The heavy plough went first and cut the furrows while the light plough followed to turn up the earth. Workers with wooden hoes would then break up the soil and sow the rows with seed. This involved carrying a basket full of seed in one hand and flinging it into the rows. To press the seed into the furrows, the livestock were driven across the fields before the workers could close them with hoes. Irrigation would then bring the water to the field as needed via the canals.

The *shaduf* was primitive labour intensive device which allowed them to raise the water from the Nile into the canals. It was a long wooden or reed tool with a bucket on one end and a weight on the other, allowing one person to lift the bucket of water and move it to another area.

Other agricultural tools included the axe, the swab, which was a piece of wood or metal attached to a stick and used for drilling and the tanbur which helped the farmer to carry water.

shaduf

Crops

Grains such as barley, emmer and einkorn wheat were grown to make bread which was a staple in the diet. Grain was cut with a sickle and taken to a threshing table. Emmer wheat spikelets had to be removed by moistening and pounding them with a pestle. It was then dried in the sun, winnowed, sieved and finally milled on a saddle quern. This coarse flour was often thickened with sand, and despite being sifted several times, most Egyptians had bad teeth from the bread consumed. Yeast was used, making Egyptian bread the oldest leavened bread in the world.

Other crops grown for food and industrial purposes:

- Legumes were staples, including beans, lentils chickpeas and fava beans.
- Vegetables grown- lettuce, parsley, onions, turnips and garlic.
- Fruit growing required more complex horticultural techniques such as propagation, cloning and basin irrigation. From the earliest days, dates, grapes, watermelon, figs and sorghum were grown, while olives, apples, pomegranate and carob were introduced during the New Kingdom.
- Herbs were used in cooking, medicine and cosmetics.
- Industrial crops like flax and papyrus were used for construction, paper making, ropes and linen clothing. Henna was grown for the production of dye.

Egyptians loved beef and bred various types of cattle such as long-horned, short-horned, polled and zebuine. They were also used as draft animals to pull ploughs. Other animals used for meat included lambs, sheep, poultry, goats and pigs in Lower Egypt. However, the best meat was reserved for the nobility who were able to hunt antelopes. Fish, such as tilapia were staples of the lower classes and caught in nets. However, they were considered unclean by many upper classes, including the priests who were forbidden to eat them. Fish was preserved by salting and drying which was vital in such a hot climate. Poultry, including duck, goose, pigeon and quail was also popular among all classes.

Painting from the burial chamber of Sennedjem

Beer, the most popular beverage, was brewed from emmer wheat. Bread and beer were the staples of the Egyptian diet. The "beer bread" was high in yeast and crumbled into vats, where it naturally fermented. This process of brewing beer was well established before the 1st Dynasty.

The oldest high-production brewery in the world was discovered in Abydos in 2021 by a joint Egyptian-American team. Dating back to the 1st Dynasty more than 5,000 years ago, it consisted of eight large areas 20-metres-long (65 ft), with each containing 40 earthenware pots arranged in two rows. A mixture of grains and water was heated in the vats, with each basin held in place by clay levers placed vertically in the form of rings. The facility was able to produce about 22,400 (5 gallons) of beer at a time. The brewery was a remarkable structure built on the cusp of Pre-dynastic and Old Kingdom times.

In 2011, the Dogfish brewery consulted ancient Egyptian recipes in order to brew a beer with traditional ingredients.

"Ta Henket is brewed with an ancient form of wheat and loaves of hearth-baked bread, and it's flavored with chamomile, doum-palm fruit and Middle Eastern herbs. To ferment this earthy ancient ale, Sam and friends traveled to Cairo, set out baited petri dishes and captured a native Egyptian saccharomyces yeast strain."

Www.dogfish.com/brewery/beer/ta-henket

Cooking in ancient Egypt was rather simple and involved a clay or stone oven, with mortar, mill, dough basin, sieves, spoons and knives for utensils. Cooking methods included barbeque, boiling, grilling, frying, drying and preserving in salt. Salt preserved foods included meats, dates and raisins. Baking bread in the Old Kingdom was often done in heavy pottery moulds and set in embers. By the New Kingdom, a large open-topped clay oven, cylindrical in shape and encased in thick mud bricks and mortar, was used. New Kingdom tomb paintings show images of bread in many shapes and sizes. Oils were derived from ben-nuts, sesame, linseed and castor. Seasonings included salt, juniper, aniseed, coriander, cumin, fennel, fenugreek and poppy seed.

The royal bakery, Ramsses III

Banquets were depicted in paintings from the Old Kingdom onwards. Seating arrangements varied according to social status, with the highest ranked sitting on chairs and the lowest ranked sitting on the floor. Professional dancers were accompanied by musicians playing lutes, tambourines and clappers. Whole roast oxen, ducks, geese and pigeons were served along with stews, in addition to large amounts of bread, vegetables and fruit. The goddess Hathor was often invoked during feasts.

Food was served at small tables three times a day, with the main meal followed by fruit or desserts such as jams or honey pies. Other flour based desserts had meat fillings.

RECIPE

Unfortunately, no ancient Egyptian recipe books survived, although numerous recipes for medicines and cosmetics have been found. The following recipe is adapted from a later Egyptian recipe with modern ingredients and methods.

Ta'amia or falafel ingredients:

- 1 lb beans soaked overnight and drained
- 2 cloves garlic, finely chopped
- 2 large onions, finely chopped
- 1-2 teaspoons each ground coriander and cumin
- 1 teaspoon cayenne pepper
- 1/2 teaspoon baking powder
- 1/4 cup minced parsley
- A pinch of salt, pepper
- Black pepper to taste
- Sesame seeds to coat the cakes
- Olive oil for frying

Method: 1. Ensure the beans are soft and remove their skins. Mix the beans together with all of the ingredients, except the oil and sesame seeds, and either mash or blend them in a food processor until you have a thick paste.

2. Set the paste aside for 30 minutes to allow the flavours to mix.

3. Knead the mixture and form into small round cakes about 2cm thick.

4. Sprinkle each side of the cakes with sesame seeds and shallow fry in hot olive oil for two to three minutes until golden brown.

5. Serve with flat bread and lettuce tossed in olive oil, lemon juice, and pepper. Alternatively serve with a tahini dip.

REFERENCE: https://ancientegyptonline.co.uk/recipes/

EGYPT, once a major exporter of grain has been unable to produce enough wheat since 2021, resulting in such high food inflation that one of the staples, a flatbread known as eish baladi had to be subsidised by the government.

Now a huge importer of grain, Egypt has suffered dramatic food insecurity since the outbreak of the Russia-Ukraine war in 2022. The world's largest importer of wheat and 10th top importer of sunflower oil has seen the prices for these products increase dramatically, virtually overnight. Unfortunately, 85% of its wheat and 73% of its sunflower oil is imported from Russia/Ukraine, so the country has been scrambling for new markets. Soybean oil is also in short supply as the chief supplier, China, is currently hoarding its soybean products for domestic use. Cooking oil has been subsidised by the government since the crisis.

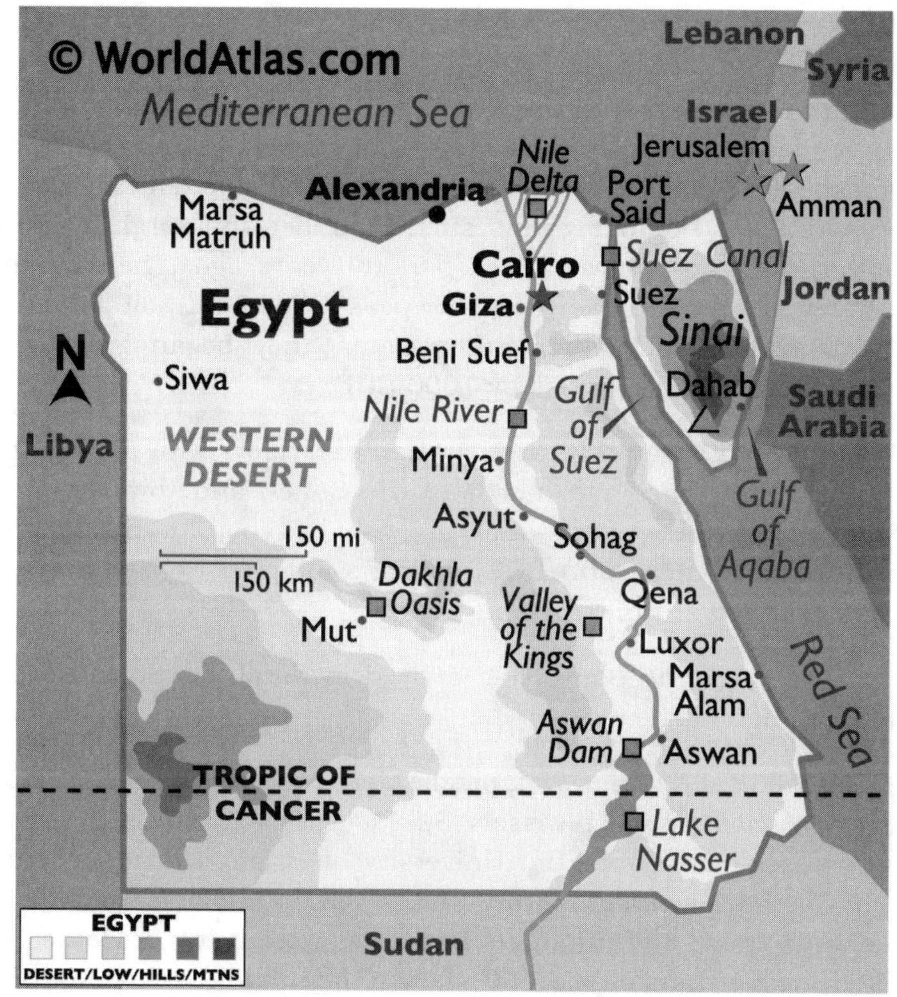

The Emergency Food Security and Resilience Support Project was launched by the World Bank in June 2022 to approve a USD500 million loan to ensure poor Egyptians have uninterrupted access to bread by expanding climate-resilient wheat silos and improve domestic cereal production.

WEST AFRICA

Farming emerged around 3000 BCE in the lush savannah on the border between Nigeria and Cameroon. Wild cattle were domesticated several thousand years before the emergence of agriculture, and fodder grains such as millet and sorghum were the earliest crops grown at least 5,000 years ago. The Africans also grew tubers such as yams, cowpeas (2500 BCE), oil palms as well as melons and other fruits. Later, they began growing a special strain of African rice by 1000 BCE.

The Bantu began a great migration of farmers with their cattle around 3,000 years ago, north to the Sahel, and towards East Africa, where a few plants such as enset (a type of banana) were domesticated. After 1000 BCE, Africans adopted iron technology which provided sturdier farming tools.

By 500 CE, the Bantu had arrived in Natal, southern Africa, via the Congo jungle.

The Central Nigerian Nok people were processing beeswax and honey in their cooking vessels 3,500 years ago according to a team of scientists from the University of Bristol and colleagues from Goethe University, Frankfurt. Their study of ancient Nok potsherds shows that products from bees were being processed or stored in one-third of the 450 sherds they studied. The Nok are known for their large-scale terracotta figurines and early iron production in West Africa. https://theconversation.com/beeswax-in-nok-pots-provides-evidence-of-early-west-african-honey-use-161197

SOUTHERN AFRICA was occupied by hunter gatherers like the San people until Bantu farmers and pastoralists moved south from Central Africa. The Khoikhoi brought herds of animals and introduced crop production and metal tools. They grew watermelons, beans and sorghum. The farmers who settled in Zimbabwe in the first millennium CE cultivated sorghum and wheat and kept cattle, sheep and goats.

ETHIOPIA

From 3000 BCE, it is likely that grain cultivation, the use of the plough and animal husbandry developed in Ethiopia along with substantial agricultural villages. Crops included teff, finger millet, ensete and coffee. Around 1000 BCE, the Ona people of Eritrea were agriculturalists cultivating native and imported crops as well. By 600 BCE, the Kingdom of Aksum emerged. Agricultural crops like wheat and barley were traded across the Indian Ocean, as was salt. By the 2nd and 3rd centuries CE, Aksum controlled the southern Red Sea basin and was the principle supplier of African goods to the Roman Empire via the revitalised Indian Ocean trading system.

Coffee beans originated in the highlands of Ethiopia. According to a legend, a goat herder Kaldi discovered coffee when he noticed how energetic his goats became after they ate berries from a certain tree. Kaldi reported his findings to the abbot of the local monastery who threw the berries into the fire as Devil's work. But the fire roasted the berries and gave off such a pleasant aroma, that the abbot poured water over them to preserve the smell. Upon drinking the mixture, he raved about how the berries kept him alert during the long hours of evening prayer.

Coffee beans were first exported to Yemen by Somali merchants in the 15th century CE and soon spread further in the Arabian Peninsula to Mecca and Medina. It rapidly spread through the Islamic world and reached Europe in the 16th century.

Other crops indigenous to Africa:

- Yams (*Dioscorea rotundata*)
- Watermelon, native to north-eastern Africa. Wild watermelon seeds were uncovered in Uan Muhuggiag in Libya dated to 3500 BCE, and were domesticated in Egypt by 2000 BCE.

- Sesame seed was domesticated over 3,000 years ago for its oil. The wild species of the genus *Sesamun* is native to sub-Saharan Africa, whereas *S. indicum*, the cultivated type, originated in India.
- Sorghum, also known as great millet, broomcorn, guinea corn, durra and jowar, originated in Africa where it was an important cereal crop for humans and animals. *Sorghum arundinaceum* was first domesticated in the Niger Valley from 7000-5000 BCE. The species developed were guinea, a West African variety requiring high rainfall, daudatum which is grown in Ethiopia and Chad and kafir, a drought-resistant type grown in Southern Africa.
- Shea nuts from Burkina Faso
- Okra from West Africa or Ethiopia
- Black eyed peas, also known as cowpeas by 2500 BCE
- Plantains were cultivated by 3000 BCE and bananas by 1000 BCE.

CENTRAL & SUBSAHARAN AFRICA from 10,000 BP showed beginnings of vegeculture, the collection of wild plants and protection of wild tubers such as yams. By regularly harvesting wild roots, the Africans developed stone hoes with a fine polish. Climate change brought the desert southward, forcing the hunters and foragers to settle down to farming, fishing and herding.

The white Guinea yam, *Dioscorea rotundata* was farmed from about 5000 BCE, enabling the population to grow in the northern savannah. During the second phase of the agricultural revolution, millet and sorghum cereals were later cultivated in the northern savannah. Eventually crops like sorghum spread to the Middle East, other parts of Africa and as far away as India.

The third stage of the food production brought an increase in tending of trees such as the oil palm which allowed for more cooking techniques to be employed. The sap of the palm tree could also make a type of wine. Eventually bananas made their way from Asia to Africa, where they were propagated by cuttings and seeds. They were probably introduced by Indian Ocean traders in their outrigger canoes. Bananas became a staple carbohydrate of many Central Africans, as they flourished in the forests.

Neolithic Central Africans were skilled fishermen using nets, and clay for making pots which allowed them to cook, brew and store food and liquids. During the Copper and Iron Ages, salt mining became an industry, as did aquaculture with fish dried for export. Dried fish could be preserved for months and carried to regions deficient in protein and sold for high prices. Fish were caught in lakes and rivers and grown in special ponds.

AFRICAN food security is at risk across the continent after several years of drought and inflation. According to the FAO, there are 33 African countries where famine is knocking on the door. The countries most affected by this situation are Sudan, Mali, Burkina Faso, Chad, Guinea, Cameroon and Niger. In the Horn of Africa, drought due to four missed rainy seasons since 2020 have placed Somalia at risk of famine, as well as reduced harvests in East African countries of Ethiopia and Kenya, the organisation warned.

The West African countries of Niger and Burkina Faso are suffering from a deficit in cereal production and high food prices. Grain prices more than doubled in the 12 months prior to August 2022 in Mali and Burkina Faso.

The World Bank has offered the following African countries loans to improve the resilience of their food systems.

- $315 million dollar loan to Chad, Ghana and Sierra Leone
- $130 million for Tunisia
- $2.3 billion for Eastern and Southern Africa

https://www.worldbank.org/en/topic/agriculture/brief/food-security-update

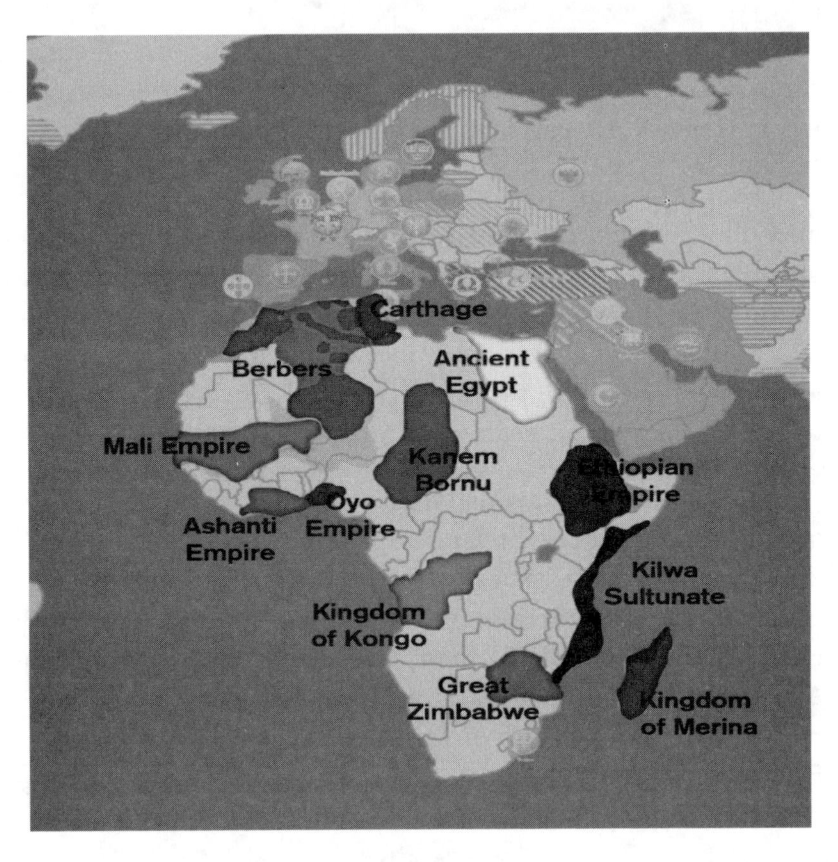

Ancient African kingdoms

PART 4

ANCIENT EUROPE

BRONZE AGE BRITONS

THE CELTS

SCANDINAVIA VIKING RECIPES

THE MINOANS

THE MYCENAEANS

ARCHAIC GREECE

CLASSICAL GREEK FOOD

GREEK POTTERY

HELLENISTIC GREECE

THE ETRUSCANS

REPUBLICAN ROMANS- RECIPE

ROMAN EMPIRE

AQUACULTURE

ROMAN FOOD TECHNOLOGIES

FOOD AS MEDICINE

ANCIENT EUROPE

This section looks at some of the important European cultures in the Bronze and Iron Ages which followed the Neolithic. Southern and Eastern European cultures entered the Bronze Age before the northern peoples of Scandinavia, Russia and the Baltic. Cultures in the Bronze Age from 2000-1000 BCE were generally pre-literate and agrarian, with the exception of the Mycenaean Greeks. The first millennium BCE saw the rise of writing, iron, complex agrarian practices and the expansion of trade under the Greeks and Romans.

BRONZE AGE BRITONS

The ancient pre-Celtic Britons lived and ate well in round-houses with cooking utensils, woven clothes and luxury items traded from Mediterranean cultures. Their diet was rich and varied. They grew cereals such as einkorn from which bread and porridge were made, as well as beans, peas and spelt.

A site known as "Must Farm" at Whittlesey in Cambridgeshire dating from 1000-800 BCE is revealing an unprecedented array of household goods, from pots with food inside them to wooden buckets and platters. Animal remains found in the rubbish dumps reveal they were eating wild boar, red deer, lambs, calves and freshwater fish such as pike.

Milk was being consumed about 6,000 years ago, but may have been fermented into cheese and yoghurt as many people were lactose intolerant.

THE CELTS were an Iron Age culture which spread from Central Europe to Britain and Ireland around 3,200 years ago. They sowed cereal crops such as emmer wheat, oats and rye. Using oxen, they ploughed the land with the iron tipped arathar, planted seeds and culled forests. In the Austrian Alps, at Hallstatt and Hallein, they mined for salt.

The Celts domesticated cattle and fenced fields as pasturelands. In Ireland they introduced herd animals such as horses and cattle, and used sheep and goats for milk, butter and their hides for clothing. Pigs were also raised.

Religion and farming were closely linked in Iron Age Britain. Granaries were often underground pits which were used to store surpluses of grain. Excavations at Danebury Iron Age Fort have revealed offerings to the gods were placed at the bottom of the pits before harvest. Ancient Greek writer Diodorus Siculus remarked,

> "In reaping their wheat they cut off the ears from the stalk, and house them in pits underground; then they take and pluck out the grains of just enough of the oldest of them to last for the day, and after they have bruised the wheat make it into bread." (Book V, chapter II)

Oats, rye and millet were introduced during the Iron Age, with emmer wheat making an appearance around 500 BCE. Barley and einkorn wheat continued to be cultivated. These grains were made into bread and porridge. Wheat was processed by setting ears alight, and extinguishing the fire when the husks were burnt. The wheat was then winnowed, baked and ground into flour with saddle querns. Few vegetables were grown in Britain before the Roman invasion, with the exception of Celtic beans, spinach, parsnip and nettles.

Land clearing was undertaken at an accelerated rate by the Belgic tribes of southern Britain in the first century BCE who used an ard with an iron ploughshare to break the soil. Pliny noted that the Celtic plough was superior to the Roman type which replaced it. Because of land depletion, the British farmers invented the practice of fertilising the soil with loam and chalk.

Cattle were prized in the Celtic world where a man's wealth was measured by the number of his herd. Cows provided milk which was turned into butter and cheese, while bullocks were slaughtered for their meat. Butchered bones found at Danebury show that sheep were butchered as mutton, not lambs, as their fleece was used for wool. Mutton constituted about a quarter of the total consumption of meat. The Hebridean and Shetland breeds are direct descendants of the ancient Iron Age sheep. The Celts kept domesticated pigs which provided most of their ham, pork and bacon. Chickens and geese were kept as pets but rarely eaten.

Meat was roasted or stewed, often in large bronze cauldrons which could be placed directly over the fire. The diners sat on the ground and ate with their fingers, cutting the meat with daggers. Celtic Gauls used salt to preserve the meat, especially pork for winter.

Shallow earthen ware pots were used as drinking vessels. Mead was brewed from honey and water left together in a pot to ferment, and then flavoured with wild herbs and fruits. The Celts also made beer from barley and wheat by allowing the grain to germinate, before heating it and leaving it to ferment. Crab apples were fermented to make alcoholic cider. They also imported wine and later began to grow vegetables.

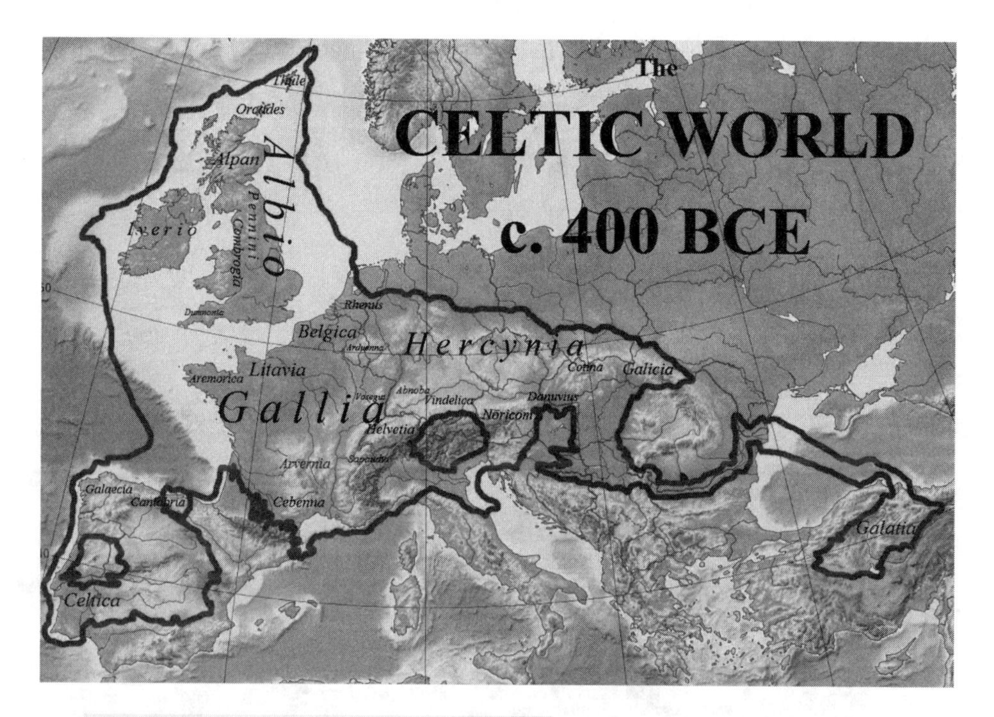

RECIPE: Irish Iron Age boar stew

Ingredients/Method: 1.Marinate overnight 2.2 lbs (1kg) cubed wild boar/pork in 3.5 cups of mead or ale, with a handful of fresh thyme, parsley, bay and 10 crushed juniper berries.

2. The following day, remove the meat from the marinade and fry in 10 tbsp of dripping or lard, adding 4 chopped carrots, ½ shredded cabbage and ¾ cup baby onions.

3. Bring reserved liquid back to the boil, adding 2 tbsps honey and ¾ cup of pearl barley.

4. After liquid has reduced by 2/3, simmer half covered for 2 hours until the sauce has thickened.

5. Serve with barley bread or griddle cakes.

6 To make griddle cakes, rub 3.5 oz of lard or butter into 9 oz spelt flour with pinch of salt. Add 1 egg and 3-6 tbsps of milk to

make a firm dough. Roll pieces into balls and flatten before placing on a hot greased griddle. Cook each side for 5 minutes.

https://www.irishcentral.com/culture/food-drink/eating-irish-iron-age-ancestors

Gundestrup Celtic cauldron from Denmark. Credit Nationalmuseet CC BY-SA 3.0

Ancient writers described large bronze cauldrons used for Celtic feasts. Dating from the Late Iron Age, the silver Gundestrup cauldron was found in pieces which had to be reassembled in Denmark.

SCANDINAVIA

The original population of Scandinavians at the end of the Ice Age were hunters and gatherers. Around 4,000 years ago new people arrived concurrently with the introduction of agriculture.

During the **Viking** era, from the 8th to 11th centuries CE, the cuisine seemed quite varied. They cultivated cereals such as barley, rye, oats and rarely wheat, in addition to raising cattle, sheep, horses, pigs, goats and poultry. Dairy was sourced from domestic cows, sheep and goats to provide milk, butter and cheese. Beans, peas, cabbage, onions and herbs were cultivated, while wild fruits such as apples, pears, cherries and berries were gathered. Fish, a staple part of the Scandinavians' diet near rivers and lakes, were caught with hooks and nets.

Vikings were skilled hunters using spears or bow and arrow. They supplemented their diet with deer, bear, boar, elk, rabbits, seals and walruses. Seabirds were harvested for their eggs and their flesh. The now extinct great auk was an important part of the diet in Iceland but was hunted to extinction.

Meat and fish were preserved by smoking or drying which were useful ways of storing food in a cold climate. Cooked meats were preserved in vats of sour whey (*súrr*) as were some dairy products.

Meat was often prepared by boiling it in a wood-lined pit. Meat and water were placed in the pit and hot stones brought the water to boiling point. This water was seasoned with herbs and vegetables. These heated stones from fire pits were also used for drying and warming food. Meat was roasted on a rotating spit or boiled in soapstone cauldrons. Some Norse homes had stone ovens which were heated by stones from the fire pit.

Bread was made from unleavened barley ground in stone querns which were used as grindstones. Dough was made from wheat, barley, rye and oats mixed with honey, whey and nuts. It was flattened and cooked on a flat pan over heated coals.

Ale, a staple beverage, was brewed from malted barley along with other drinks like milk, beer and mead. Barley grain was soaked in water and allowed to grow shoots to create malt which was dried, crushed and heated in hot water for fermentation. It was allowed to ferment in an iron cauldron or wooden vat.

Norse families ate two meals a day: *dagverðr* at mid-morning, and *náttverðr* in the evening. Meals were served on wooden trenchers and eaten with a knife, while stews and porridge were served in wooden bowls and eaten with a spoon. Beverages were consumed with wooden cups or drinking horns at feasts. Copper, silver or glass vessels were also used by wealthier people.

REFERENCE: Hurstwic: Food, Diet, and Nutrition in the Viking Age

http://www.hurstwic.org/history/articles/daily_living/text/food_and_diet.htm

Norse farmers in the Viking age practised Landnam (land take) and shipped large numbers of grazing livestock to colonies in Iceland and Greenland. These animals included cattle, sheep, goats, pigs and horses which grazed from May to September and housed in individual farms in winter. The farmers also felled the trees, drained the bogs and cut peat to irrigate their fields.

The volcanic soils in Iceland and Greenland are low in clay and high in organics. The removal of the bogs by the Vikings caused the introduced Scandinavian plant species to squeeze out the local plants. The Norse extensively manured the soils in the first few years, but environmental degradation and climate change in the Medieval Little Ice age from 1100-1300 CE forced the Greenland colonies to be abandoned. Much of the Iceland topsoil was removed, leading to massive erosion which still afflicts the country.

RECIPE: Kornmjölsgröt (Barley Porridge) Adapted from a later recipe.

Ingredients:

- 10-15 cups of water
- salt
- Two cups of chopped barley kernels, soaked overnight in cold water
- A handful whole grain wheat flour
- A handful crushed hazelnuts
- 3-4 tablespoons of honey

Method:

Put the ingredients in a large pot. Pour 10 cups of water in the kettle and heat to a rolling boil. Stir regularly, reducing heat if needed to maintain a low boil. Add water if needed if the mixture starts getting too thick. Cook until done. This takes me about an hour and makes about 4 to 6 servings.

REFERENCE: Traditional Viking Foods
thevikingworld / Traditional Viking Foods (pbworks.com)

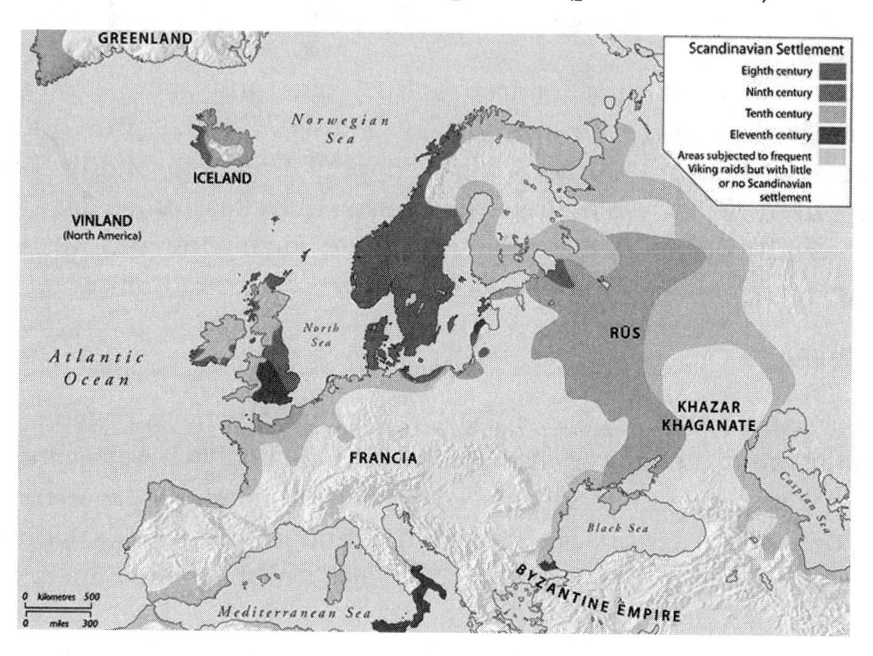

ANCIENT GREECE

As Greece is a hilly and rocky country with many islands, fishing has always been a part of the economy as well as grapes and olives.

CRETE

The oldest Greek civilisation known as the Minoan culture in Crete had its zenith before 1450 BCE when the Palace of Knossos was built. We know from archeology that the Cretans had sophisticated water technologies for domestic and agricultural use. The major agricultural goods traded were olives, figs, grains, wine and saffron. They were also skilled mariners who traded with other cultures around the Mediterranean.

Olive oil was widely used in Crete for food, beauty and ointments. The olive press in Vathypetro, Archanes is believed to be the oldest in Europe. The palace of Kato Zagros yielded a preserved olive in a water cistern. Huge stone urns called pithoi which stored the oil were found at the Palace of Knossos, providing it with a massive storage capacity.

The Minoans raised sheep, cattle, pigs and goats and grew barley, chickpeas, grapes, figs, lentils, vetch and olives. Pomegranates and quinces were adopted from the Middle East. They used wooden ploughs with leather handles pulled by oxen, along with bronze double axes and adzes for hoeing and weeding. Simple wood sickles were used for cutting wheat and barley.

Food

In 2014, American archeologist Jerolyn Morrison organised a symposium held at the American Farm School where attendees could eat ancient Cretan food cooked in replicas of the utensils. On the menu: lentils with coriander and honey, snails sautéed with rosemary and black pig fillet with carob cooked in a clay replica of a Minoan grill. He said,

“There are many foods that we eat today, that they used to eat in Minoan Crete. During both the Minoan era, and in modern Crete, people consume food from the sea, such as small and large fish, barnacles, cuttlefish and sea snails. They also ate meat from animals they bred or hunted such as goat, sheep, hare, pigs and cattle, while in later Minoan periods they also ate wild deer. Excavations have brought to light the fact that they used to eat many legumes, such as lentils and beans, cereals such as barley and zea, fruits and nuts, such as figs and almonds and of course there was evidence for producing olive oil and wine! I believe that the Cretan diet from 100 years ago was more like that of the Minoan era than modern Greece.”

REFERENCE: Zikakou, Ioanna, ‘The Food of Ancient Cretans,” “Greek Reporter,” June 6. 2014.

https://greekreporter.com/2014/06/06/the-food-of-ancient-cretans/

THE MYCENAEANS were the earliest culture in mainland Greece, from 1600 to 1200 BCE. Their economy was agrarian, based mainly upon barley and wheat. They also grew grapes and figs which were used to make wine and olive oil. Other indigenous crops were carrots and pears. Archeological evidence shows that the walnut, chestnut and rye were introduced into Mycenaean culture by 1400 BCE. Goats, pigs, sheep and cattle were herded, while hunting was a past time of the upper classes.

Tablets written in Linear B, the language of the Mycenaeans, record offerings to the gods, particularly the goddess Potiniya, in appeasement for a good harvest. These offerings include wine, honey, sacrificial rams, goats or pigs.

These tablets also describe spices which may have been imported, such as coriander and cumin. Excavations of the palace at Pylos have yielded food preparation equipment such as souvlaki trays, tripod cooking pots and ovens. These souvlaki trays are large rectangular trays which most likely held skewers for grilling meat. Linear B tablets found at Pylos reveal that the palace employed hundreds of specialised food preparation workers, including 90 grain grinders and 10 "mixers."

The Mycenaeans were master engineers and builders who created great stone fortresses such as at Mycenae and Pylos. Less well known is their expertise in hydraulic engineering, such as draining Lake Kopaïs to increase arable and pasture land over 3,000 years ago. Archeologists estimate that about 70 million cubic feet of earth was excavated to create canals and embankments, with the water diverted away from the lake into the Bay of Larymna.

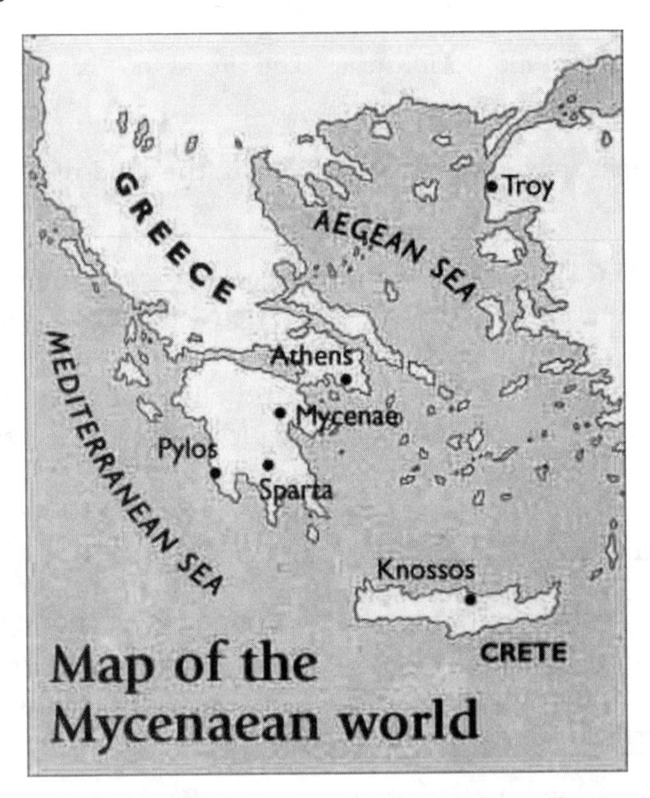

Map of the Mycenaean world

ARCHAIC GREECE 1200-550 BCE This period covers the collapse of the Mycenaean Empire into the dark ages and expansive Greek colonisation around the Mediterranean into areas with arable farming like Sicily, Spain, North Africa and the

Black Sea. Fortunately, three pieces of literature have survived from this period: Homer's "Iliad" and "Odyssey" and Hesiod's "Works and Days" which provided much information about farming during the dark ages.

Farming Throughout ancient Greek history agriculture was practised by a large percentage of the population. The main cereal crops were barley and emmer wheat. Olive trees were suited to the hilly, rocky terrain, and olive oil was one of the main exports of many separate city states known as *poleis.* Other vegetables grown were cabbage, onion, lentils, chick peas and beans with herbs like sage, thyme, mint and oregano. Fruit such as fig, apple and pear trees were grown in orchards, while almonds, sesame, linseed and poppy were also grown for their oils.

Due to the rugged terrain, goats and sheep became the most common livestock, with beef reserved for the upper classes. These animals provided meat, hides, wool and milk in the form of cheese. Poultry such as chickens and geese were domesticated along with pigs. Beekeeping provided honey which was used in medicines and the production of mead. Sugar was unknown.

Hesiod's 8th century BCE "Works and Days" and Xenophon's "Economy" (4th century BCE) provide much information about agricultural crops such as olives.

The olives were harvested from late autumn to early winter either by hand or a pole. Placed in wicker baskets, they were left to ferment for a few weeks before being pressed. The oil was preserved in terracotta vases. Trees and vines were pruned and legumes were harvested.

For other crops:

- During spring which was the rainy season, the farmers brought fallow land into production, as they practised biennial crop rotation, alternating from year to year between cultivated and fallow.

- In summer, irrigation was needed. In June, the crops were harvested with a sickle. Oxen were used to trample the wheat for threshing and the grain was stored, while women and slaves ground it for bread making.
- In early autumn, the farmers used a wooden ard to break the hard crust on the soil, as well as a hoe and mallet. The farmers sowed the fallow land for the following year by hand. The grapes were harvested, crushed by foot in large vats and the wine was left to ferment in jugs.

By Xenophon's day four centuries later, little advancement had been made in agriculture, and the tools remained basic.

In **Athens**, the Eupatrides class were the landed gentry whose farms were worked by peasants. The great lawgiver Solon introduced measures to help the peasantry, including abolishing slavery for debt in 594 BCE. The Spartans under Lycurgus had dramatic redistribution of land with 10 to 18 hectare lots (kleroi) allocated to each citizen.

In **Sparta**, the diet was far more basic. Black soup, made from boiled pork with salt, vinegar and blood was favoured by the military class. Bread was not the staple it was in other Greek poleis, and meat from hunting was popular, along with dairy products like milk and fruits. Watered down wine was also consumed, but never to the point of drunkedness.

During Archaic times, different Greek states set up overseas colonies for settlement, trade and agriculture. They founded colonies in Italy, the Adriatic and Black Sea where the land was more arable. In return for cereal and products like timber and metals, the Greeks exported their famous pottery.

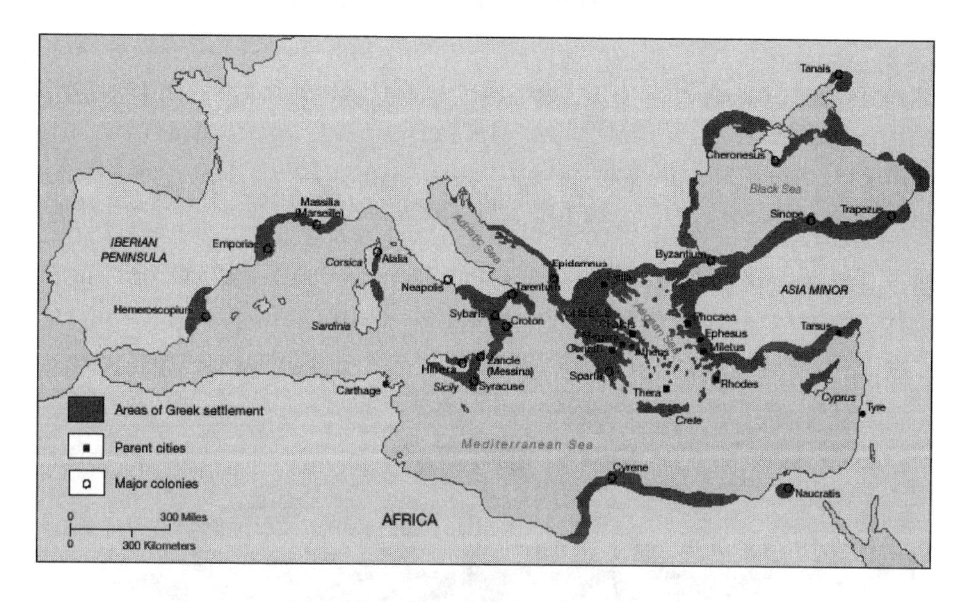

Greek colonisation

CLASSICAL GREEK FOOD 550-325 BCE

The Greeks had three to four meals a day, starting with breakfast (ἀκρατισμός) which consisted of barley bread dipped in wine with figs or olives. They also ate a pancake called τηγανίτης (*tēganítēs*), made with wheat flour, olive oil and curdled milk which was pan fried. A different pancake was made of spelt flour and topped with honey, cheese and sesame seeds.

A quick lunch (ἄριστον *áriston*) was taken around noon and consisted of figs, salted fish, bread, cheese and olives. "Lunch dinner" was often served in the late afternoon. Dinner (δεῖπνον deĩpnon) was served at nightfall and was the most important meal of the day. Women and men ate separately, with the men eating first. Meals consisted of bread made from wheat or barley, baked at home and accompanied by cheese and honey. Barley flour was used to make maza, which could be cooked or raw, made into dumplings or flatbread. Legumes such as black beans, broad beans, lentils, chickpeas, peas and lupin beans were common foods. Vegetables such as cabbage, carrots, cucumbers and asparagus were eaten as soups or boiled and mashed with olive oil, vinegar, herbs, or garon, a fish sauce. Meat and seafood consumption varied in accordance with the wealth and location of the household. Pork was the cheapest meat to buy in the marketplace, followed by kid and lamb. Even the peasants had chicken, geese and eggs for consumption.

Fruits were eaten as dessert, along with nuts. Dried or fresh, the most important fruits were figs, raisins and pomegranates.

Greek plate from the Louvre 350-325 BCE

Seafood was a staple of the Greek islands and coastal regions. They harvested squid, octopus, sardines, mackerel, anchovies, tuna and carp. Lake Kopaïs was famous for its eels; in Athens conger eels were considered a great delicacy.

Wine was the most important drink in ancient Greece, including red, white, rose and port wines. All wine was cut with water, as it was considered barbaric to drink it straight. They also drank *kykeon*, a combination of barley gruel, water, herbs and goat cheese.

Dairy products and cheese making were widespread in Hesiod's day. Goat and ewe cheese was a staple food, along with curdled milk products similar to yoghurt. Cheese was also used as an ingredient in cooking, including fish dishes.

Mithaecus was a cookbook author from Sicily in the late 5th BCE. His is the first known Greek cookbook, but only one brief recipe survives thanks to a quotation in the Deipnosophistae of Athenaeus. The recipe for cooking the cepola fish calls to "Gut, discard the head, rinse, slice add cheese and oil."

Archestratus in the 4th century BCE titled himself as the "inventor of made dishes" and describes a recipe for pauch and tripe cooked in "cumin juice, and vinegar and sharp, strong-smelling silphium."

(Wikipedia: Ancient Greek Cuisine")

A famous Athenian cultural institution of the Classical age was the **symposium,** a main meal in the evening when men laid around on couches indulging in wine and discussion, often accompanied by naked flute girls. The most famous example is Plato's "Symposium" which depicts a group of men, including Socrates, attending a banquet and engaging in philosophical questions about love and desire.

“Plato’s Symposium” depiction by Anselim Feuerbach

GREEK POTTERY was first made in Crete and continued into Mycenaean and Classical ages. It was made from clay which was orange-red in Attica and pale buff in Corinth on the potter’s wheel in separate sections which were joined together. Then the pots were decorated with paint before being fired at around 960 degrees C in kilns. Often they were fired three times in order to enhance the painting.

For centuries, potters created wares for practical use such as holding wine, water and perfumes as well as cups and plates. These include:

- Kraters for mixing wine with water
- Oinochoai for pouring wine
- Kylixes for drinking- a cup with two handles
- Hydra for holding water with three handles
- Skyphoi deep bowls
- Lekythoi jars for holding oils and perfumes
- Amphora were large vessels used for storage.

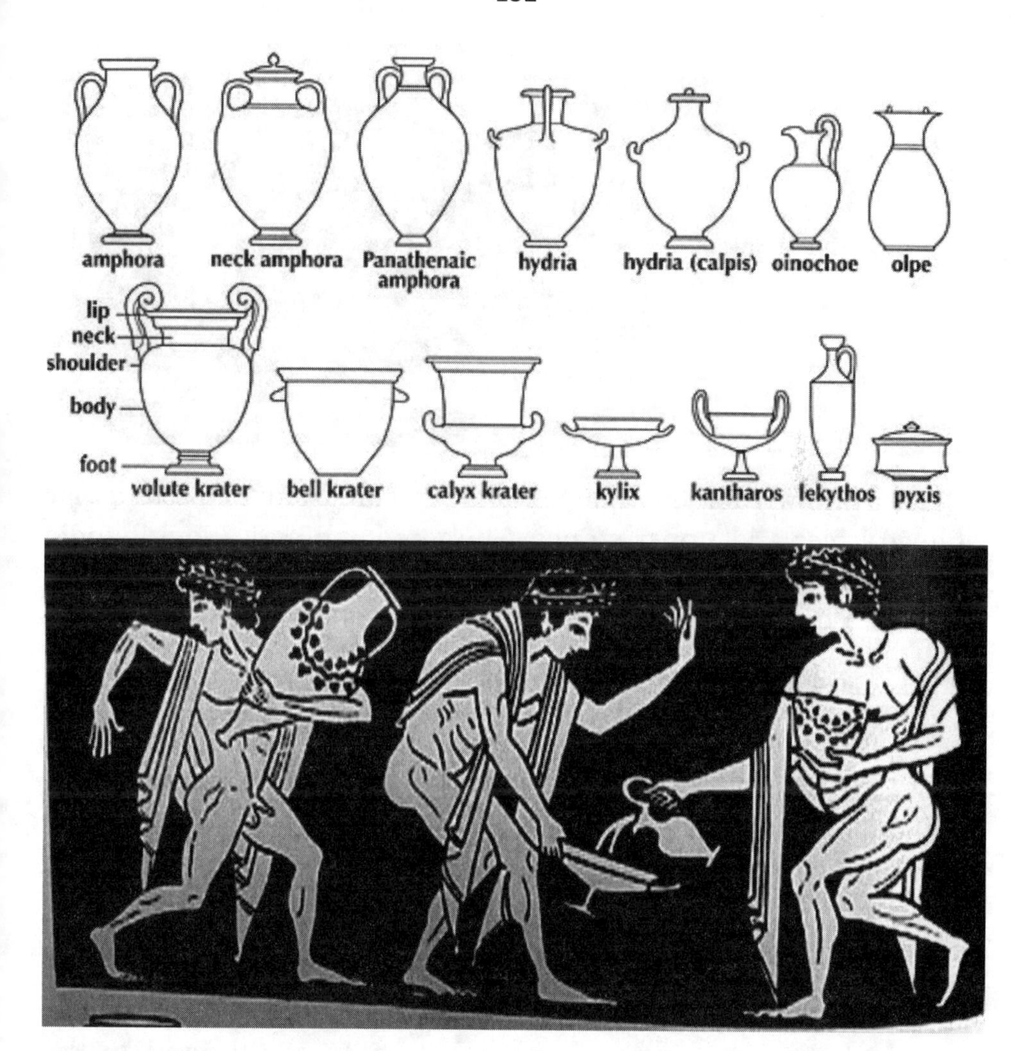

Images: https://www.kenneymencher.com/2014/03/art-history-everyone-should-know-greek.html

The Olynthus Mill was a device developed to grind grain in the late 5th century BCE.

Also known as the hopper mill, it consisted of rectangular shaped lower and upper stones. The top stone had a long handle which was pulled from side to side while the grain was fed through a narrow slot at the centre.

Greek Agricultural Mythology

The Greek term for the gods of agriculture was "Theoi Georgioi" or "Theoi Khthonioi" which translated to "gods of the earth." The major deities were Demeter, the Olympian goddess of agriculture, including wheat and barley as well as flour and bread, her daughter Persephone, queen of the underworld and goddess of grain, and Dionysus, god of viticulture and wine.

The story of Demeter and her beloved daughter Persephone explained the changing seasons and fertility of the land to the Greeks. Persephone was also the daughter of the great god Zeus, whose brother Hades of the underworld desired her. Zeus arranged for Persephone to wander away from her mother and guardian nymphs into a garden, where Hades emerged from the underworld with his chariot of black horses and abducted her.

Demeter was utterly distraught at the abduction of her daughter and neglected her duties, so that agriculture wilted and famine ravaged the earth. When Zeus heard of the calamity, he sent Hermes to the underworld to retrieve Persephone, but was surprised to find her living as a queen in a beautiful garden.

Although Persephone had learned to love Hades, she also missed her mother. Hades gifted her with six pomegranate seeds to eat, ensuring that she would be compelled to return.

The rape of Persephone vase

Upon meeting with Zeus, she told him she would like to stay with her husband, but Demeter was so enraged that she threatened to stop attending to the crops. Zeus then decided that Persephone was to spend six months with her mother and six months with Hades every year. During the time she spent with Hades, Demeter would cause the earth to wither and die, which became autumn and winter. Her reunification with her mother represented fertility and fruitfulness in spring and summer.

The goddesses Demeter and Persephone most likely predate Greek civilisation and may have had Cretan origins. They were also worshipped in Mystery Religions such as the Eleusinian Mysteries.

The Eleusinian Mysteries were annual initiations for the agrarian cult of Demeter and Persephone based at the Panhellenic Sanctuary of Eleusis near Athens. They commemorated the myth of the abduction of Persephone, Demeter's search for her daughter and the ascent of Persephone to be reunited with her mother. The original story was written down in the 7th century BCE by an anonymous author in the Homeric Hymns. The cult itself survived into the late Roman Empire.

The rites and ceremonies were kept secret, but historians have made a good effort to understand them. The rebirth of Persephone symbolised the eternity of plant life, and initiates believed in achieving a reward in the afterlife. The high point of the ceremony was "an ear of grain cut in silence."

The participants included priests, priestesses, initiates, others who had participated at least once, and those who had learned the secrets of the greatest mysteries of Demeter. There were Lesser and Greater Mysteries: the Lesser Mysteries took place in the eight month of the Attic calendar around February/March when participants would sacrifice a piglet to Demeter and Persephone before ritually purifying themselves in the river Illisos. The Greater Mysteries took place around September/October and involved bringing sacred objects from Eleusis to the Eleusinion temple in Athens. Other sacred rites included an animal sacrifice, ritual bathing, a great feast and finally a procession of people walking to Eleusis along the Sacred Way. After an all-night vigil, initiates entered a great hall called Telesterion where rituals were held, including a secret climax, and finally a great feast.

The Dionysian Mysteries were associated with the god of wine and viticulture, including the grapevine's cultivation, fermentation and intoxicating qualities. Dionysus (Bacchus to the Romans) was the god of revelry, orchards, fruit, vegetation, ritual madness and religious ecstasy. His parents were believed to be Zeus and Persephone and he was Thracian by birth.

The rites were based on a seasonal death and rebirth theme, common among other agrarian cults like the Eleusinian Mysteries. The consumption of wine and other substances, as well as dancing, music and shouting induced a state of ecstasy or even trance in the participants. During the trance, the spirit of Dionysus himself took possession of the body. Euripides, the Athenian playwright, described Dionysian rites in "The Bacchae."

> "Following the torches as they dipped and swayed in the darkness, they climbed mountain paths with head thrown back and eyes glazed, dancing to the beat of the drum which stirred their blood' [or 'staggered drunkenly with what was known as the Dionysus gait']. 'In this state of ekstasis or enthusiasmos, they abandoned themselves, dancing wildly and shouting 'Euoi!' [the god's name] and at that moment of intense rapture became identified with the god himself. They became filled with his spirit and acquired divine powers."

Both men and women were inducted into the Dionysian Mysteries which involved four stages:

The desire to join and apply
A preparation period
The sacred rites were performed
One is integrated with other initiates.

Recipes

This recipe from Roman Cato dates to about 200 BCE and includes instructions for Greek olive relish.

> "How to make green, black or mixed olive relish. Remove stones from green, black or mixed olives, then prepare as follows: chop them and add oil, vinegar, coriander, cumin, fennel, rue, mint. Pot them: the oil should cover them. Ready to use."– Cato, On Agriculture 119

Athenian cabbage:

"Cabbage should be sliced with the sharpest possible iron blade, then washed, drained, and chopped with plenty of coriander and rue. Then sprinkle with honey vinegar and add just a little bit of silphium. Incidentally, you can eat this as a meze.' – *Mnesitheus, quoted in Oribasius, Medical Collections 4, 4, 1*

A modern adaptation of the recipe:

Ingredients

- 1 small white cabbage
- 2 heaped tsp chopped fresh green coriander in oil
- 2 tsp chopped fresh or dried rue (you can use a bitter herb or spice such as fenugreek seed as a substitute)
- 2 pinches asafoetida powder (you can use garlic or onion powder as a substitute)
- Salt

Honey vinegar

- 120g honey
- 2 tbsp red wine vinegar

Method

First, make the honey vinegar. Boil the honey and skim it, add the vinegar and reduce a little. Store until needed. Finely slice the cabbage, wash and drain it. Toss with the herbs and 3 tablespoons of honey vinegar and sprinkle with the asafoetida powder and a little salt. https://blog.britishmuseum.org/cook-a-classical-feast-nine-recipes-from-ancient-greece-and-rome/

REFERENCE: 'Cook a classic feast: nine recipes from ancient Greece and Rome," The British Museum, https://blog.britishmuseum.org/cook-a-classical-feast-nine-recipes-from-ancient-greece-and-rome/

Hippocrates was a Greek physician born in 460 BCE on the island of Cos who is attributed with the aphorism "let food be thy medicine." He is commonly referred to as the Father of Medicine who revolutionised ancient Greek medicine by separating it from religion and establishing it as a profession. Hippocrates is also credited with applying the theory of "humours" the vital body fluids of blood, phlegm, yellow bile and black bile to medicine. These humours needed to be balanced for a body to be in a state of health and were produced as part of the digestion process.

Hippocratic medicine was based on the healing power of nature, as the body itself has the power to rebalance the four humours and heal itself. Treatments involved rest, cleanliness of wounds, fasting and the consumption of honey mixed with vinegar. He believed that fevers and other diseases were treated by fasting.

The Hippocratic Corpus is a collection of seventy early medical texts collected in Hellenistic Greece written by various authors. Food was defined as more forceful or less forceful. To avoid a diet that was too forceful, it had to be mixed and cooked. Excessive consumption of wine caused disturbances of thought as well as stomach problems of pain, diarrhoea and vomiting.

HELLENISTIC EMPIRES

The Macedonian Empire was forged by Alexander the Great in 323 BCE and continued under his generals Ptolemy and Seleucus, as well as various dynastic kingdoms which lasted until 31 BCE with the Roman conquest of Egypt. At its height, it covered a vast area from Egypt in the south to Persia and western India. The arts and science flourished during this Hellenistic age, as new trading routes with Asia and Africa brought many fresh ideas and exotic foods to Greece for the first time. Science bloomed, with great minds such as Archimedes who invented or copied innovative methods for irrigation.

The trading routes brought many herbs, spices and fruits from Asia to the Hellenistic world. Pepper and ginger were introduced, as well as other Indian spices like nigella, spikenard and asafoetida. Cinnamon was quickly embraced as a flavouring. Rice from China was imported as a medicine rather than a food and not grown in the west.

Alexander's expeditions brought botanists, such as Theophrastus, along to identify viable species, and the most useful fruit import was citrus which was believed to be native to tropical Asian areas. This ancient citrus called citron (*Citrus medica*) was a large fragrant lemon shaped fruit with a thick rind which was first recorded by Theophrastus in 350 CE.

Theophrastus was a student of Aristotle in Athens and presided over the Peripatetic school for thirty-six years. Apart from philosophy and ethics, he was considered the father of botany, and spent time studying plants in Lesbos while his colleague Aristotle studied animals. His two large botanical treatises are "Enquiry into Plants" (*Historia Plantarum)* and "On the Causes of Plants."

The "Enquiry of Plants," originally ten volumes, classifies plants into their modes of generation, their sizes as well as their localities and practical uses such as juices, herbs and foods.

- Book 2 deals with reproduction of plants as well as how and when to sow them.
- Book 7 deals with herbs, book 8 describes plants with edible seeds.
- Book 9 concentrates upon plants that produce juices, gums and resins.

"On the Causes of Plants" was originally eight books, of which six survive. These books cover the growth of plants, when to sow and reap various crops, preparing soils, using tools and the economical properties of many plants.

Originally Alexander's tutor, he accompanied him to the east, and was the first westerner to describe the cotton-plant, banyan, pepper, cinnamon, myrrh and frankincense.

AGRICULTURE AND IRRIGATION

Agriculture spread throughout the empire, with grain growing areas in Egypt, Cyprus and Phoenicia under control of the Ptolemaic branch, and Mesopotamia under control of the Seleucid branch, providing wheat. The Ptolemies of Egypt dominated agriculture by renting farming tools, instituting an official planting schedule and collecting all produce to sell to the population. Olives were cultivated on state-controlled lands by peasants, extracted by contracted labour and sold by licensed dealers at fixed prices. The Ptolemaic government taxed imported olive oil at 50% to protect its own state run business.

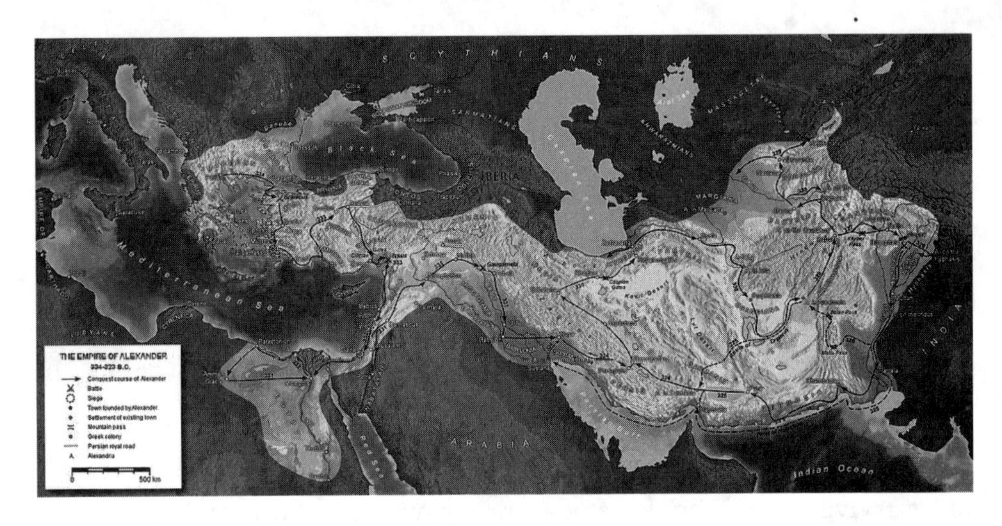

Alexander's Empire- credit Generic Mapping Tools CC BY-SA 3.0

The Seleucid Empire which was founded by Seleucus 1 Nictator, existed from 312 to 63 BCE. At its height it covered Anatolia, Persia, the Levant, Afghanistan and parts of Turkmenistan. Around 80-90% of the population was involved in agriculture, food production and trade. Agricultural structures including temples, royal estates and poleis (cities) were usually established and inherited from the Neo-Babylonians and Achaemenid predecessors. Private property did not exist; any land not delegated to the temples or poleis was considered the sovereign's private property. The main products of Mesopotamian temple land consisted of dates, barley, mustard, sesame, cardamom and wool.

The Seleucids also maintained the ancient Mesopotamian and Persian waterways. As agriculture was a great source of state income, the kings actively managed the irrigation and reclamation projects. The Pallacottas canal built to control the level of the Euphrates took more than two months to build by more than 10,000 Assyrians.

Inventions Water wheels use flowing water to create power by means of paddles mounted around the wheel. As the water moves the paddles, the rotation of the wheel is transmitted to machinery such as a grinding millstone via the shaft of the wheel. Although water wheels were used in ancient Mesopotamia, the earliest descriptions of them are found in Hellenistic sources. Apollonius of Perge described vertical watermills in around 240 BCE, but his work only survives today in Arabic translation. Antipater of Thessalonika made reference to a water wheel which was probably a vertical wheel. Philo of Byzantium (280-220 BCE) described wheels for submerging siege mines.

The wheels were used for crop irrigation and grinding grains, which had formerly been done by humans and draft animals. The first wheels were horizontal, whereby water flowed from a stream or aqueduct, thus turning the wheel forward which rotated the grindstone. These horizontal wheels with a vertical axis called a Norse mill, Greek mill or tub wheel were a primitive type of turbine. Later, the Greeks used vertical water wheels.

The earliest depiction of a compartmented wheel is from a tomb painting in Ptolemaic Egypt dating from the 2nd century BCE which shows a **sakia** gear. This gear uses buckets attached to a vertical wheel which in turn is attached by a drive shaft to a horizontal wheel pulled by oxen.

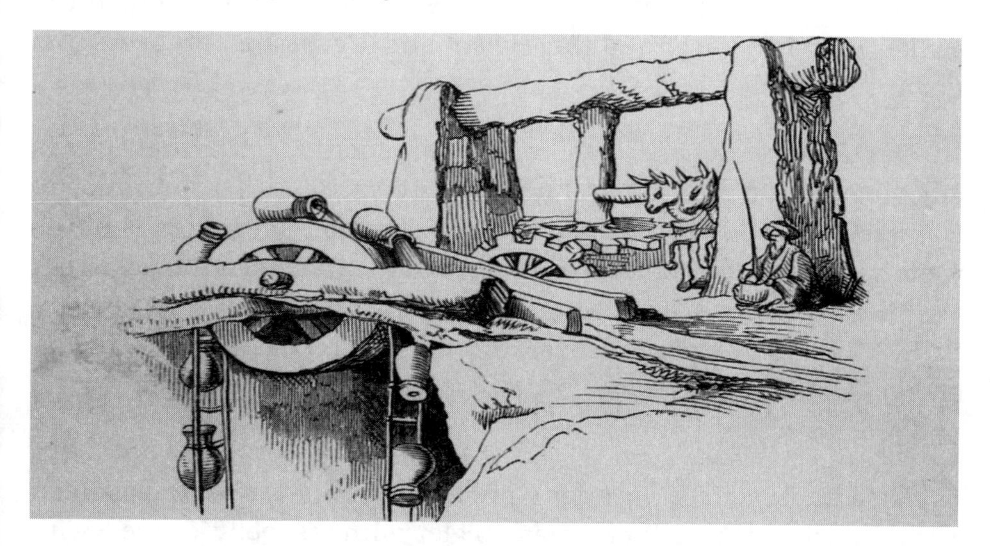

Sakia gear, credit interestingengineering.com

Archimedes was a famous philosopher and inventor. Whilst visiting Egypt around 234 BCE he described a water screw which could transfer water from a river or lake into irrigation ditches at a different level. Also known as the Egyptian screw, the **Archimedes screw** consisted of tubes would round a cylinder which rotated, lifting the water within the spiral tube to a higher elevation.

However, it is possible that the screw was invented by the Babylonians, as the author Strabo described the Hanging Gardens which were irrigated by screws. Whether Archimedes invented the device or copied it from the Hellenistic Egyptians, it bears his name.

The Hellenistic age was one of prosperity, with the Greeks enjoying a cosmopolitan and luxurious lifestyle hitherto unknown. Banquets in the Persian style became common, with details of the wedding banquet of Macedonian Caranous preserved for posterity. Caranous's banquet for twenty men was early in the third century BCE. Guests received silver bowls, and feasted on bread, chickens, ducks, pigeons, geese, turtledoves, partridges, hares, and kids as a first course.

The main course, brought on a silver platter with a golden edge, consisted of a large roast piglet with its belly stuffed with roasted thrushes, fig-peckers, oysters and scallops with egg yolks spread on top. Other dishes included hot kid, roast fish, boars, Cappadocian bread (white and slightly sour) and unlimited wine.

Dessert was a simpler affair, with flat cakes and Cretan gastrin.

Recipe: Myma was a meat dish of the 1st century BCE which included giblets, cheese, blood and 13 herbs and spices. Epainetos says in his "Art of Cookery" that "A myma of any sacrificial animal, or chicken, is to be made like this: chop the lean meat finely, chop liver and offal finely, mince with blood and flavor with vinegar, baked cheese, silphium, cumin, fresh and dried thyme, Roman hyssop, coriander leaf, coriander seed, Welsh onion, fried onion or poppy seed, raisins or honey and the seeds of sour pomegranate. You may also serve it as a relish." (1st.cent.BC, apud Athen. 662d)

REFERENCE, Kavroulaki, Mariana, "History of Greek Food,"

https://1historyofgreekfood.wordpress.com/2010/12/03/going-hellenistic/

Banquet scene from the tomb of Agios Athanasios, Thessaloniki, 4th century BCE

ANCIENT ITALY

The Villanovians, an Italian Iron Age culture, now identified as early Etruscan, were influenced by the Greeks. The **Etruscans** culture lasted from around 900 BCE until they were conquered by the Romans and incorporated into their empire in 27 BCE. Etruscan territory reached its maximum around 750 BCE, covering Etruria, the Po Valley and Campania. Although their writing was borrowed from Greek, it is only partially translated today, so we have to rely on archeology and the Roman writers like Livy who lived centuries after the Etruscans were subsumed into the Roman Empire.

Mario Lopes Pegna, the Italian Etruscologist, praised the Etruscans for their methods of land improvement and water management. Their skilfully constructed network of canals collected surplus water and channelled it to farms. The Etruscans first developed the technique of dry farming and applied it to the arid soils of the Maremma hills. Dry farming is crop production without irrigation using methods such as early soil preparation, planting drought tolerant cultivars, diligent weed control and other methods to improve the water holding capacity of the soil.

Many areas were rich in soils, and the Etruscans were also skilled at draining swamps and reclaiming areas for crops. Agricultural production was dominated by Mediterranean polyculture of wheat, vine and olive, while sheep, cattle and pigs were the basis of animal husbandry. Sheep and goats were important in the later Bronze Age, while cattle gained prominence for consumption by the main Iron Age culture. Chickens were imported in the early Iron Age and may have been used for ritualistic purposes, before being popular farmyard animals by the 4th century BCE.

Mediterranean crops like olives, vines, fruits and wheat were restricted to the lower valleys and basins. *'Coltura promiscua'* was a polyculture of growing olives, vines and cereals to provide water control. Transhumance, moving of flocks or herds of animals between lowland winter and upland summer pastures, was practised.

Wine was fermented during the Iron Age, and became an important component of feasting rituals and daily consumption. Many bronze and ceramic pots found in Etruscan tombs relate to storage, mixing, pouring and drinking wine.

Etruscan tombs also reveal pictorial images of feasts which show that roasting was a main method of cooking. The ceramic cooking stand (*fornello*) was developed in the Bronze Age and allowed vessels to cook food quickly or slowly, allowing the production of stews, gruels and dairy such as cheese and yoghurt.

Land improvement schemes such as draining marshland indicated the sophisticated hydraulic engineering skills of the Etruscans. Around 625 BCE, low lying marshy areas in Rome were cleared by drainage channels, and the stream separating the Capitoline and Palantine had its embankments strengthened and was eventually covered. The Cloaca Maxima, or great drain, was built around this time in Rome.

In Viterbo, Veii and Todi, the Etruscans built innumerable channels, or *cuniculi* in the volcanic tufa to drain the waters. The delta area of the Po River was prone to catastrophic flooding in prehistoric times before the Etruscans dug a network of canals and lagoons while damming the river with caissons or brushwood. The ancient Etruscan city of Spina was discovered in the delta area in 1922, after it had been flooded for many centuries.

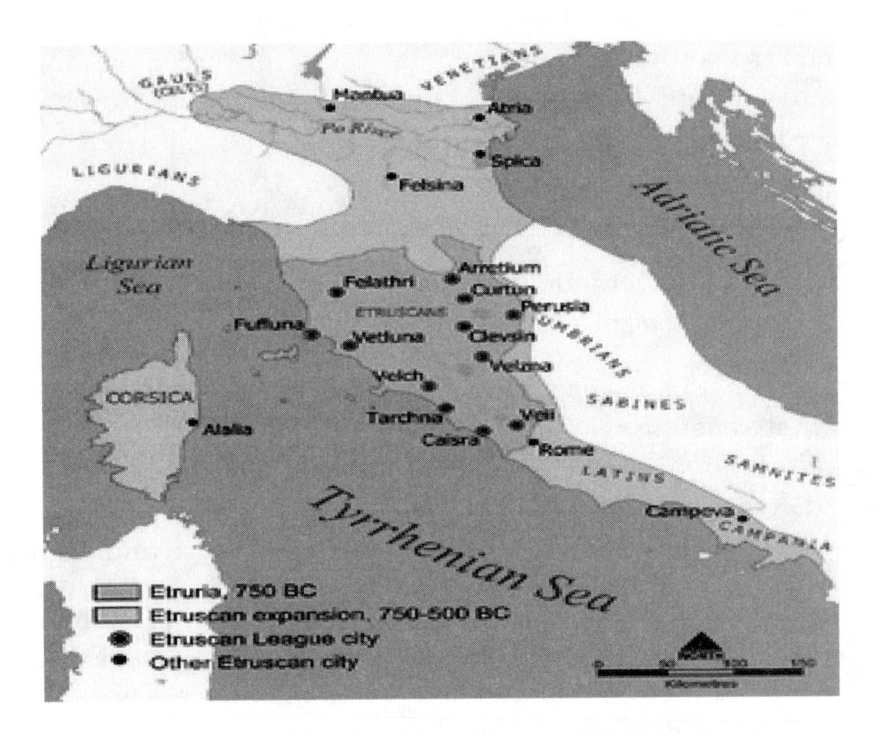

Map Norman Einstein CC BY-SA

Banquet scene, Tomb of the Leopards

Cuisine

The Etruscans contributed much to Roman cooking. Murals in tombs show images of banquets with metal saucepans filled with vegetables next to meat sizzling on a grill over the fire. They were believed to be amongst the first people to make and use a form of pasta in many dishes. Etruscan wine, called *'sassicaia'* was very potent and had to be watered down. Olive oil was a major export, and the Etruscans were the first to cultivate the olive on an industrial scale for exportation. Cattle, pigs and sheep were raised for meat, with wild boar a particular favourite. Steaks were cut up and either grilled over a fire or cooked in a thick stew for the banquet table.

Journalist John Hooper attended an Etruscan banquet at the Castello di Potentino in Tuscany in 2019. He described the menu "chickpeas with cumin, fennel and defrutum, a light fruit syrup; seasoned fish with asparagus; rough-ground pork sausages smoked in hay and goat's-cheese cakes flavoured with honey and lavender. Overall, the flavours are more redolent of contemporary Middle Eastern cooking than of today's Italian fare."

Hooper, John, 'How to eat like an Etruscan did 2,000 years ago," 'The Economist,' November 14, 2019.

https://www.economist.com/1843/2019/11/14/how-to-eat-like-an-etruscan-did-2000-years-ago

Aruspicini is a very old recipe inspired by the Etruscan 'aruspice' divination priest who analysed animal entrails to divine the future. These are unleavened spelt buns stuffed with pork liver, tongue and green sauce, or goose. Measurements of ingredients were not included.

Recipe

- Mix spelt flour and water until it reaches the consistency of pizza dough and cook in a pan.
- Prepare a vegetable soup and introduce the tongue when it's hot.
- After cooking remove the tongue and discard the skin.
- Cut into thin slices when cooled.
- Green sauce- parsley leaves, garlic, capers and anchovies in vinegar.

'Tastes from the Past: The Etruscan recipes'

https://www.visittuscany.com/en/ideas/tastes-from-the-past-the-etruscan-recipes/

REPUBLICAN ROME traditionally existed from 508 to 27 BCE, when the Roman Empire was born. From the 2nd century BCE, the Punic and other Wars had spread Roman hegemony around the Mediterranean, opening new trade routes and colonies. The Republican Romans looked upon the Etruscans as licentious and extravagant, attributes frowned upon, as their traditional ideals were more directed towards virtue, dignity, gravity and simplicity. Their food was also simple, but became more luxurious under influences from the Hellenistic Greeks and Carthaginians whom they had conquered.

Republican meals consisted of *jentaculum* (breakfast), the *cena* (lunch) and the *vesperna* (dinner.) Breakfast was usually a piece of bread with honey or cheese, or puls, a porridge made of emmer, salt olive oil and herbs. *Cena* was the main meal of the day, while *vesperna* consisted of a light supper.

Roman food, British Museum CC BY 2.0

During the late Empire, the *cena* extended later into the afternoon, with two or three courses, and eventually the *vesperna* disappeared.

Farming was the backbone of the Republican economy, and was considered a noble profession. Cicero declared in *"On Duties"* that "of all the occupations by which gain is secured, none is better than agriculture, none more profitable, none more delightful, none more becoming to a free man." (Cicero, "On Duties," 75)

- Grains such as wheat, barley and millet contributed to about 70% of the Roman diet with emmer, spelt and durum the most prominent forms of wheat. Durum became popular in the cities because it could be baked into leavened bread. Other foods grown and consumed:
- Vegetables: artichokes, mustard, celery, parsnip, beets, asparagus, cucumbers, gourds, cabbage, and fennel, olives.
- Fruits like plums, figs apricots and peaches.
- Herbs like mint, rue, basil, coriander, rocket chives, dill, capers, saffron, parsley, marjoram and cumin
- Dairy in the form of cow, sheep and goat's milk which was used to make cheese
- Nuts included almonds, walnuts and chestnuts.
- Poultry, including chickens, ducks and geese provided food and eggs.
- Most common exports were olive oil and wine which was the preferred beverage. In the 2nd century BCE large estates called **latifundia** were growing these crops on an industrial scale by employing slaves.
- Honey was the only sweetener.

Autumn and spring agricultural calendar:

- September, ploughing for grain crops and sowing a month later. Grape and olive harvests also came in the autumn months.
- During spring months other crops were sown; later in cooler areas.

Grains were grown in Sicily, Sardinia and North Africa which became a Roman province after the 3rd Punic War destroyed Carthage. North Africa and Sicily eventually became granaries for the Roman Empire.

Cato the Elder, a Roman censor, wrote "*De agri cultura*" in 160 BCE, which is the oldest surviving work of Latin prose. Although criticised for its rambling prose, it was more like a manual of farming intended for friends and neighbours. The work contains much information on planting and caring of vineyards, including how to treat the slaves who maintained them. This was the time when the latifundia were becoming larger and more prosperous, as many of these new estates were sixty acres in area, requiring a large working force of slaves.

Another Republican Roman, Marcus Terentius **Varro** wrote his comprehensive treatise "*Rerum rusticarum libri tres*" (Three books on Agriculture) in the final years of the Republic. Dedicated to his wife Fundania, the first book was on agriculture, the second book on cattle and the third on game and fish preserves.

From the introduction of book II, he wrote,

> "Each of these three classes has two stages: the earlier, which the frugality of the ancients observed, and the latter, which modern luxury has now added. For instance, first came the ancient stage of our ancestors, in which there were simply two aviaries: the barn-yard on the ground in which the hens fed — and their returns were eggs and chickens — and the other above ground, in which were the pigeons, either in cotes or on the roof of the villa. On the other hand, in these days, the aviaries have changed their name and have become ornithones; and those which the dainty palate of the owner has constructed have larger buildings for the sheltering of fieldfares and peafowl than whole villas used to have in those days."

REFERENCE: Varro, *"De Re Rustica,"* Loeb Classical Library, 1934.

Several Roman inventions made harvesting more efficient. A new type of grain sickle eased the strain on the user's wrist, while a large scythe was designed for mowing tall grasses. A new harvesting machine, invented by the Trevires Gauls, had metal teeth at the front of a wooden frame on wheels. The teeth cut off the grains which fell into the wooden frame behind pushed by mules or oxen. Shovels made of iron helped in the preparation of the soil before planting vines and fruit trees. Iron hoes, triangular in shape, were also important farming tools.

Gallo-Roman harvester WikiCommons

Water for farms was often brought in by aqueducts which carried water from the hills to the fields and towns. The Romans also dug trenches to drain wetlands, and irrigated fields.

RECIPES

Cato's book had recipes for farm products such as Coan wine, *vinum Graecum, savillum, libum* and *placenta,* cheesecake like pastries. Although variations of placenta had been used since Classical Greek times, Cato wrote down this recipe:

> Shape the *placenta* as follows: place a single row of *tracta* (dough) along the whole length of the base dough. This is then covered with the mixture [cheese and honey] from the mortar. Place another row of *tracta* on top and go on doing so

> until all the cheese and honey have been used up. Finish with a layer of *tracta*...place the placenta in the oven and put a preheated lid on top of it [...] When ready, honey is poured over the placenta. (Cato the Elder, *De Agri Cultura*)

Variations of placenta cake are still enjoyed in Greece as baklava as well as throughout the Balkans and Romania.

Agricultural reforms

During the 2nd century BCE, Rome's population was growing rapidly, and developing trade routes across the Mediterranean helped bring grain to the capital. Wheat and barley were the primary cereal crops, with barley being the cheapest to produce for the poor. Unfortunately, it was also the most difficult to mill, so the impoverished could not afford to bake bread but simply ate it as porridge or flat cakes.

Grains were measured by the *modius,* (2.4 gallons.) Authors like Cato, Sallust and Seneca recorded that every person in Rome was provided with four *modii* of grain per month. However, Rome, unlike other contemporary cities such as Alexandria and Carthage, had to import most of its grain from Sicily, Sardinia and North Africa, leaving it vulnerable to food shortages.

Although there had been land reform in earlier centuries of the Republic, the Gracchi brothers, Tiberius and Gauis Gracchus, initiated sweeping land reforms which facilitated more grain imports into Rome. As Tribune of the Plebs, Tiberius brought in reforms that dealt with equitable land distribution after the capture of enemy territory. His agrarian law sought to allot this land to the poor, including landless soldiers returning from active duty. These reforms attacked the wealthy landowners and politicians who were acquiring land for their latifundia. In 133 BCE, he was murdered by the aristocrats.

Gaius Gracchus was then elected Tribune and initiated further agrarian reforms such as the *Lex frumentaria* which provided a monthly distribution of grain to Roman citizens at a fixed cost, avoiding price fluctuations. Unfortunately, Gaius also fell foul of the ruling class and was driven to suicide in 123 BCE.

Tiberius and Gaius Gracchus, Musee d'Orsay, Paris

Over the next century there were many adjustments to the system of grain distribution, by politicians including Pompey, who was given total control of the grain supply of the Roman world in 57 BCE, after defeating the pirates who had menaced the trade routes. Julius Caesar created *aediles cereals,* officials to oversee the grain distribution, the market and trade. Augustus assigned magistrates to the charge of grain distribution, and when famine hit Rome, the people petitioned him to become dictator and take control of the grain supply. Augustus ruled efficiently, appointing officials to manage the grain supply, and issuing grain during times of famine. It was during the dictatorship of Augustus that the Roman Republic transitioned into the Roman Empire.

THE ROMAN EMPIRE: 27 BCE- 476 CE

The first two centuries, known as *Pax romana,* witnessed unprecedented growth and stability of the Roman Empire, which reached its territorial peak during the reign of Trajan (98-117 CE.) The Empire extended:

- West to Spain, France and England
- East to the Black Sea regions, Greece, Asia Minor, Armenia, the Levant and Mesopotamia
- South to Egypt and North Africa

Many of these areas such as Sicily, Spain, Egypt, North Africa and the Black Sea provinces had fertile soils, and were able to provide enough grain for imperial needs. However, due to poor harvests, supply chain disruptions and civil disturbances in the provinces, there were often shortages of grain in Rome.

Under Augustus, about 320,000 people were eligible for free grain, although this stabilised at about 200,000 recipients who were issued with a ticket of entitlement. The distribution of free grain remained until the end of the Empire, although this was replaced by baked bread in the 3rd century CE. Later emperors often provided free oil, pork and wine. Eventually other cities of the Empire, such as Alexandria, Constantinople and Antioch, also began providing similar benefits.

Lucius Junius Moderatus Columella was born in Spain during the early years of the Empire. His "*De re rustica*" in twelve volumes is the most important extant book on agriculture during Imperial times. The books focus upon soils; viticulture; fruits; olive trees; cattle, horses mules; asses, sheep, goats and pigs; fish and fowl, chicken, doves, thrushes, peacocks, guinea fowl, geese, ducks, fish ponds; wild animals, bee-keeping and production of honey and wax; gardens; calendars and household management.

Latifundia The wealthy class owned luxurious country or seaside villas as well as large industrial estates producing wine or olive oil for export. By imperial times, many small holdings had been absorbed into the vast estates, often owned by an absent landlord who lived in the city. Many nobles had latifundia in the provinces as well, which subsidised their lifestyle in Rome.

Slaves who had been captured from the defeated territories laboured on these vast estates, subject to the authority of a steward (*vilicus)* or superintendent. Grain was no longer widely grown as it was imported from the territories, but vineyards and olive groves were the most profitable. Most estates had a small vineyard as well as a grape press which was used after the slaves first stomped on the grapes in large vats.

Cattle were widely bred, mainly for draft purposes or for milk, and more rarely, beef. Donkeys and mules pulled carts and carried heavy loads. Cheese from cows, sheep and goats was made in large quantities, as butter was widely unknown except in the Celtic province of Gaul. Pork was the favoured meat dish of the Romans. Sheep were kept mainly for the wool, although the wealthy enjoyed eating lamb. Bee keeping was also an important industry, because honey was the only sweetener apart from fruit.

Birds and fowl were kept in large numbers on farms for their meat, feathers and eggs. As pigeons were a common dish, dovecotes were built to keep them during the cooler months. Chickens, ducks and geese were also bred on the farms.

Olive oil had multiple uses and was stored in large amphora jars buried in the ground. Not only was it used for cooking, but also for soap and as fuel for lamps. Excess oil was exported or sold at the market.

Other small scale industries were necessary on these large estates. Enough grain was grown which was ground in the farm mills and baked by slave bakers. The mill wheel was usually turned by a horse or mule. Wool was shorn, carded, spun and woven into cloth by female slaves who were in charge of providing clothes for the workers. Male slaves were used as builders, smiths who made the tools, and carpenters, as well as ploughmen and reapers. They also picked the grapes and trod on them to create wines.

CUISINE

During the Empire, Rome became more prosperous and food became more diverse. Romans became acquainted with the crops, foods and cooking methods of the provinces. The Silk Road trade routes between China and Europe, which opened up during the late Republic, brought foods and spices from Asia to wealthy Roman households.

- Bread made from wheat was mainly reserved for the well-to-do, while the poor folk ate barley bread. Emmer, einkorn and spelt were the main varieties of wheat, while millet and oats were considered less desirable. White loaves were leavened by wild yeasts and sourdough cultures. The Spanish and Gaulish Celts used brewer's yeast to make their bread rise.
- Legumes included lentils, chickpeas, peas and broad beans. They were planted in rotation with cereals to enrich the soil, and stockpiled in case of famine.
- Pottage could be enhanced with vegetables, meat, cheese or herbs to produce an early type of risotto or polenta. Julian

stew, which was eaten by the soldiers of Julius Caesar, was made from spelt which was enriched with ground meat, pepper, fennel, wine and hard bread.

- Vegetables like leafy greens and herbs were eaten as salads with oil and vinegar dressings. The Romans grew over 20 kinds of vegetables. Beets, leeks carrots and gourds were cooked and prepared with sauces. Truffles and mushrooms were foraged, as were snails.
- Exotic fruits from the East were propagated throughout the Roman Empire, including cherries from the Black Sea, Turkey, apricots from Armenia, plums from Syria and the pomegranates from North Africa. Romans also ate berries, currants, dates, peaches, quinces, melons, figs, apples, pears and citrons.
- Nuts eaten included almonds, hazelnuts, pine nuts, pistachios, chestnuts and walnuts. Roman farmers became adept at grafting fruit and nut trees.
- Butchers sold fresh meat, including pork, mutton or lamb, and nothing went to waste as blood puddings and stews contained entrails and eyeballs. Farmers cured ham and bacon and made sausages with ground meat, herbs, nuts and egg as a binding agent. They were then aged in a smoker.
- Olive oil was a staple for cooking and lamps. The Romans invented the **trapetum** for extracting olive oil. This device consisted of a large stone bowl into which olives were poured and then crushed under two concave stones attached to a central beam fixed to an iron pivot. The apparatus then slotted onto a central post set into the bowl which allowed the stones to be turned inside it. Massive lever presses were developed for extracting larger olives in North Africa. Spain was also a major exporter of oil, including Liburnian Oil, which was flavoured with herbs and salt.

Reconstruction of a trapetum, credit Heinz-Josef Lucking, CC

- Salt became an indispensible seasoning and preserving product during the Roman Empire. As it was imported, pure salt was rather expensive, but *garum*, fermented fish sauce, was very popular. Pepper, imported from the East was so popular that pepper pots were created to hold it. Local spices included cumin, coriander, juniper berries, while saffron, cinnamon and asafoetida were all imported from the East. Carob was added to dishes for its chocolate like flavour.
- Seafood was popular in coastal regions. Mullet was especially prized. Shellfish, oysters, mussels and sea urchins were farmed at Baiae near Naples.
- Wine was consumed with meals and was offered as a libation to the household gods. Most Italian areas produced their own wines, as did southern Gaul and Spain. Wines could be black, red, white or yellow.

Imperial Romans ate three meals a day: *jentaculum* (breakfast) *prandium* (lunch) and *cena* or dinner was the main meal.

Prandium replaced the cena as a light lunch, and the cena became a multi-course meal. However, the lower classes still kept lunch as the main meal.

Wealthy Romans threw lavish feasts which lasted for hours. In contrast to the Greek symposium which was primarily a drinking party, the *convivium* was focussed on food. Both men and women reclined on couches to eat. The feast consisted of three courses: *gustation*, (appetizer) *mensae primae* (main) and *mensae secundae* (dessert.) The most lavish banquets included exotic produce which reflected the host's wealth and affluence, such as peacock tongues, roasted ostrich, songbirds, lobster, wild boar and pheasant.

A cookbook "*De re culinaria*" attributed to **Apicius**, is thought to have been compiled in the 1st century CE. The Latin text is organised into ten books, each one devoted to particular ingredients. Book 2- ground beef, 3- vegetables, 4 mixed ingredients, 5- legumes, 6- poultry, 7- gourmet, 8– four legged animals, 9- seafood, 10- the fisherman.

The recipes are geared towards the wealthy classes and provide an intimate look at their banquet cuisine. Such recipes and ingredients include:

1. Boiled ostrich in a sauce of pepper, mint, cumin, celery seeds, dates, honey, vinegar, raisin wine, garum and oil thickened with flour. This dish contains ingredients from far-flung corners of the Roman Empire, such as African ostrich, Indian pepper, Middle Eastern dates and garum from Spain.

2. Hot kid or lamb stew. Place meat pieces into the pan. Finely chop onion, coriander, pound pepper, lovage, cumin, garum, oil and win and place in a shallow pan before thickening with wheat starch. (Wikipedia, Apicius)

Other Roman recipes include stuffed dormice, sow's womb, flamingo tongues, roasted songbirds and snails.

This menu for a sumptuous banquet was described by Apicius:

Appetiser:

- Jellyfish and eggs
- Sow's udders stuffed with salted sea urchins and brains cooked with milk and eggs, boiled tree fungi with peppered fish sauce and sea urchins with spices, honey and egg sauce.

Main courses:

- Fallow deer roasted with onion sauce, rue, Jericho dates, raisins, oil and honey
- Boiled ostrich with sweet sauce turtle dove boiled in its feathers
- Roasted parrot
- Dormice stuffed with pork and pine kernels
- Flamingo boiled with dates
- Ham boiled with figs and bay leaves, rubbed with honey, baked in a pastry crust

Desserts:

- Fricassee of roses with pastry
- Stoned dates stuffed with uts and pine kernels, fried in honey
- Hot African sweet-wine cakes with honey

REFERENCE: Anderson, Marge, "Menu for a Roman Banquet," "Big Site of History," May 19, 2008. https://bigsiteofhistory.com/menu-for-a-roman-banquet/

The complexity of these recipes show that Roman chefs were proficient in baking, roasting, boiling and frying foods, using such utensils as knives, ladles spoons, pots, metal pans and cauldrons, mostly made out of metal or clay. The Romans ate many dishes with their hands until the fork was developed in the 3rd century CE. Common people used wooden spoons, cups and dishes, whereas the wealthy could afford bronze. Spoons were used for sauces, dessert, soups or stews and knives were often made of iron. Glass wear was adopted from the Hellenistic Greeks, who created intensely coloured drinking vessels. In the 1st century CE, the glass industry underwent technical growth, with the introduction of glass blowing and dominance of clear glasses. Often imported from centres like Alexandria, glass was becoming more common across the Roman Empire by the beginning of the 2nd century CE.

Family feast fresco from Pompei

The common folk ate a diet based around bread and supplemented with legumes such as chickpeas, dried fruits, cheese, olive oil and meat offal. Many lived in multi-storeyed *insula* apartments with no cooking facilities, so street food was very popular. The *taberna* was a fixed food stall with an oven serving takeaways. Diners were able to eat at long tables seated on benches in a *thermopolium,* similar to a modern restaurant. At least 150 *thermopolium* were identified in Pompeii alone. Street vendors were popular and often plied their trade in bath houses.

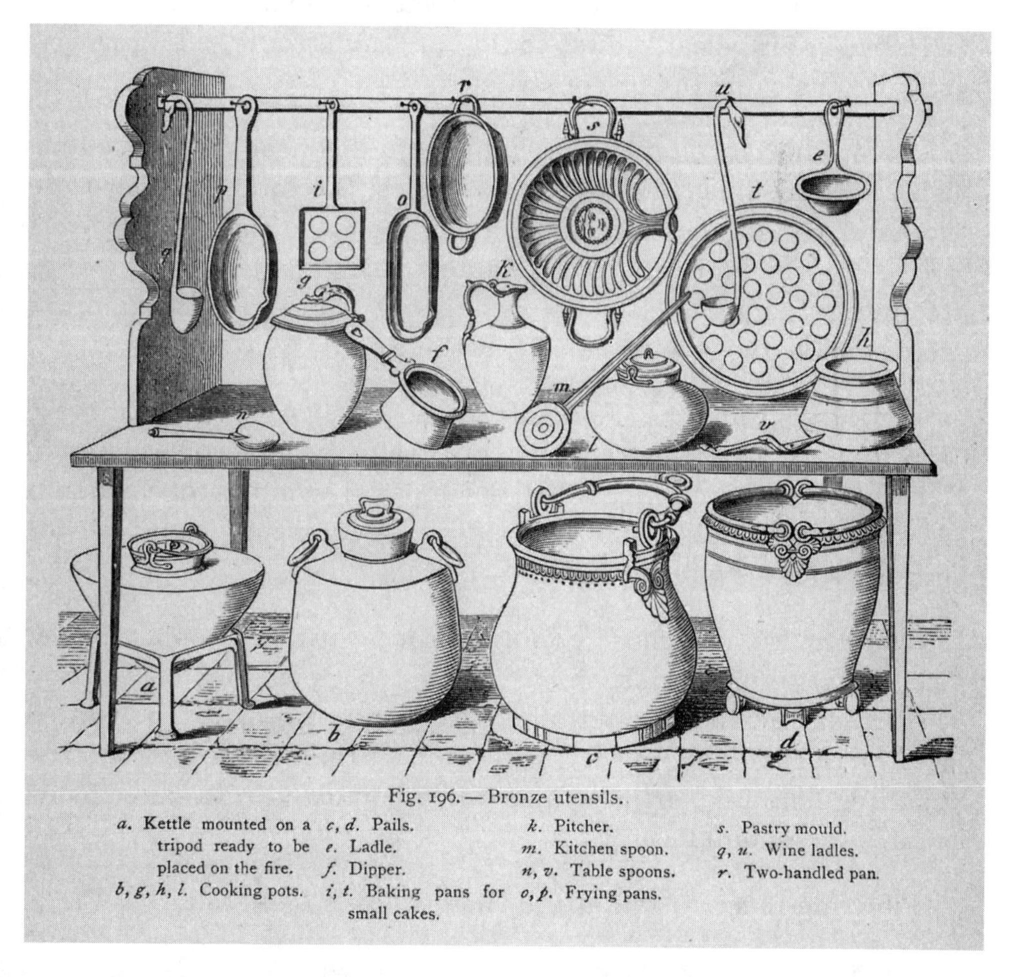

Fig. 196. — Bronze utensils.

a. Kettle mounted on a tripod ready to be placed on the fire.
b, g, h, l. Cooking pots.
c, d. Pails.
e. Ladle.
f. Dipper.
i, t. Baking pans for small cakes.
k. Pitcher.
m. Kitchen spoon.
n, v. Table spoons.
o, p. Frying pans.
s. Pastry mould.
q, u. Wine ladles.
r. Two-handled pan.

Roman cooking utensils

AQUACULTURE

Large artificial fish enclosures were known in Sumeria, Babylonia and Etruscan Italy where oysters and fish were farmed in Mediterranean lagoons from 500 BCE. Pliny the Elder credits L.

Licinius with the invention of the *piscina* in the early 1st century BCE. Varro wrote about the hedonistic practices of the *piscinarii* around 37 BCE. A century later Columella provided a more practical manual of fish keeping in his "*De re rustica.*"

Originally constructed from stones, piscinae were eventually built from hydraulic concrete, made with pozzolana from the Bay of Naples. This volcanic dust could cure underwater, allowing vast offshore enclosures to be constructed.

According to Columella, breams were amongst the first edible fish raised in Baiae, in the Bay of Naples. Sole and flounder were common pond fishes and popular as food and pets. Mullet was prized by the Romans and adapted easily to the brackish or freshwater ponds. Eels, such as moray, congers and lampreys were some of the favourite pond dwellers.

Starting from the Republican era, saltwater ponds were preferable to freshwater ponds, but from the time of Augustus, freshwater fish were regarded as imperial symbols of virtue. It became fashionable to incorporate freshwater ponds with fountains and lavish gardens.

Garum was the most popular sauce eaten in Roman times. According to Pliny "It is prepared from the intestines of fish and various parts which would otherwise be thrown away, macerated in salt; so that it is, in fact, the result of their putrefaction." The resulting liquid is garum, which when combined with other foods creates the umani flavour.

—Pliny the Elder, "The Natural History," vols I and II.

Reconstruction of Roman fish-salting plant at Neapolis.

ROMAN FOOD TECHNOLOGIES

Grain mills The animal-driven rotary mill appeared in Italy around the 3rd century BCE and was an improvement on the Olynthus Mill which relied on manpower. These mills were found in Pompeii and were driven by two donkeys harnessed to a wooden frame. The mill consisted of two parts; the lower stone which remained immobile and the upper concave *catillus* stone which was turned by donkeys and had a hole in the middle through which the miller poured the grain.

The **watermill** appeared in the last century BCE. Built next to a river or aqueduct, water striking the paddles on the waterwheel drove the upper *catillus* stone with more power than animals could provide. Watermills could produce even greater amounts of flour, and were associated with the rise of commercial bakeries. Sometimes channels, reservoirs and sluice gates were part of the mill complex, so that the water admitted to the paddle wheels could be controlled.

Another innovation was the conversion of the horizontal-wheel to the vertical-wheeled watermill through the use of a right-angled gear around 270 BCE. This gear consisted of two cogwheels which increased the velocity of the *catillus,* generating additional power. Three variants of the vertical-wheeled mill included:

- the undershot mill (water hitting the bottom of the wheel)
- the overshot (water hitting the paddles at the top of the wheel)
- the breastshot water wheel (water hitting the middle of the wheel)

The undershot was the most common application, because of its simplicity.

When populations exceeded about 500, clusters of flour mills were built, such as the Janiculum (3rd century CE) or the Barbegal complex in Gaul (early 2nd century CE.) This industrial complex, discovered by Fernand Benoit in 1940, was impressive, with 16 giant water wheels driving the massive millstones, so that the water from the aqueduct ran down the steep hill, driving the wheels one at a time. The capacity of this mill has been estimated at 4.5 tons of flour per day, enough to feed the inhabitants of Arelate.

Model of the water mills at Barbegal in Musée de l'Arles antique. Carole Raddato CC BY SA 2.0

Olive and wine presses In the 2nd century BCE, Cato the Elder wrote about presses and how to build a press room in *De Agri Cultura.* This lever or beam press was built on an elevated platform and consisted of a large horizontal beam held up by two upright fixtures beneath which pressure was applied to the grapes by a rope_fixed_windlass. The juice fell through a shallow basin towards the exit.

In the 1st century CE, Pliny the Elder described a "Greek style" press in his "*Natural History*" that had a vertical screw in place of the windlass. Varro and Columella also described similar presses in their agricultural treatises. By the 2nd century CE, Romans used a “screw press” which would include a large beam with a hole cut out of the middle through which a screw was fitted. A huge stone, which was attached to the beam by ropes and pulleys, was lifted by about eight slaves who would then walk around the screw clockwise, lowering the stone to the grapes. The juice would drain into amphorae or other large fermentation vessels.

Smaller farms used slaves to stomp on the grapes, but in large estates, presses were huge, up to 12 metres long, 3-4 metres high and weighing many tonnes. The wooden beams holding the stone were operated by ropes and pulleys built by the rural population using local stone and wood.

Olive presses were similar. Heavy stone slabs were lowered onto olives that had already been crushed in a device like the trapetum. Apart from presses, oil complexes included milling stones, decantation basins and storage vessels like amphorae. Other pressing processes were called *mola molearia, canallis et solea, torcular, prelum and tudicula.* Such machines used levers and counterweights to increase pressure on the olives to extract as much oil as possible. These presses were able to generate about 50 gallons (200 litres) of oil and 120 gal (450 l) of amurca per ton of olives. Amurca, left over water from the milling process, was often discarded, but sometimes used for grease and to treat wounds. It was also used to seal olive jars, make plaster and improve the burning of firewood.

During the Roman era, up to 30 million litres (8 mill gall) of oil per year were produced in Tripolitania, Libya, and 40 mill litres (10.5 mill gall) in Byzacena. Industrial oileries in the province of Spain produced 5 to 26 million gallons (20-100 mill litres) of oil annually. Archeological investigations at Monte Testaccio suggest that Rome imported about 6.5 billion litres of olive oil over the period of 260 years.

REFERENCE: “The Ancient History of Making Olive Oil, “ThoughtCo,” https://www.thoughtco.com/ancient-history-of-making-olive-oil-4047748

FOOD AS MEDICINE

Galen of Pergamum was a Greek physician who lived in the Roman Empire from 129-216 CE. Considered to be the most accomplished medical researcher of antiquity, he wrote about anatomy, physiology, pathology, pharmacology, neurology, philosophy and logic. Well-travelled, he eventually settled in Rome and became personal physician to several emperors.

He further developed the Hippocratic theory of the four humours: black bile, yellow bile, blood and phlegm, which dominated European medical science for over 1,300 years. These humours influenced human moods and health. An imbalance of each humour corresponded with a particular human temperament:

- blood- sanguine (social and extroverted)
- black bile- melancholic (creative, kind and helpful)
- yellow bile- choleric (energetic, passionate, prone to anger)
- phlegm- phlegmatic (dependable and affectionate)

Galen believed sickness came from an imbalance of the humours, and the best way to treat such imbalances was by bloodletting, enemas, vomiting and food. His book, "*De alimentorum facultatibus*" translates as "On the Properties of Foodstuffs," and advises correct nutrition for humour imbalances. An individual's humoural makeup determined the diseases they could get, their character and emotional state as well as the diet they should follow. However, there were no prescribed nutritional guidelines, so a diagnosis and tailor made regimen for every patient was considered crucial. Two examples of foodstuffs:

- Chickpeas are nutritious, less prone to flatulence and serve as an aphrodisiac. One recipe recommended overnight soak chickpeas, then simmer for an hour, adding salt, oil and oregano. When fully cooked, take a dry cheese, pound it to the consistency of flour, sprinkle on top of the chickpeas and serve.

- Peaches should not be eaten after a meal because they float on the surface of the stomach where they corrupt. However, when eaten at the beginning of a meal, they serve as a lubricant, helping other foods down the digestive tract.

REFERENCE, Albala, Ken, “Food: A Culinary History,” The Great Courses.

Engraved by Georg Paul Busch, Public Domain

PART 5

ASIA AND THE PACIFIC ISLANDS

INDUS VALLEY—RECIPE

VEDIC INDIA

GUPTA INDIA

SRI LANKA

INDONESIA—RECIPE

SPICE ISLAND TRADE

THAILAND

VIETNAM

CAMBODIA

NEOLITHIC CHINA

CHINESE AGRICULTURAL MYTHS

BRONZE & IRON AGE AGRICULTURE

CHINESE CROPS

CUISINE—RECIPE

KOREA

JAPAN—RECIPE

PAPUA & NEW GUINEA

THE LAPITA

HAWAII

AUSTRALIA

INDUS VALLEY CIVILISATION 3300-1300 BCE

This Bronze Age Civilisation, also known as the Harappan, which flourished in the Indus Valley, Pakistan and India, was noted for its advanced water systems, urban planning and metallurgy. The earliest known settlement was at Mehrgarh, a Neolithic mountain site which flourished from 7000 BCE, where the inhabitants domesticated barley, oats, wheat and herd animals like zebu cattle.

The early Harappan phase began around 3300 BCE when farmers from the mountains moved towards the lowland valleys. Trade networks developed and crops such as peas, sesame seeds, dates, cotton were grown, and animals like the water buffalo were domesticated. By the end of the early Harappan period large walled settlements were built, pottery industries were established and trade networks were expanded.

During the Mature Harappan, the river flood supported farming led to large agricultural surpluses. By 2600 BCE, large urban centres like Harappa, Mohenjo-daro, Lothal and Dholavira were thriving. These people built advanced hydraulic complexes such as urban sanitation systems (sewage and water supply) and docks at Lothal for shipping. They also practised dual winter/summer cropping.

Agriculture

A ploughed field at Kalibangan shows that the plough was in use by the 3rd millennium BCE. Criss-cross furrows allowed two crops to be raised in the same field. Rabi crops are sown in November or December and harvested in April or May. Harappan rabi cropping included wheat, such as emmer, shot and einkorn, all of which made bread. Three to four types of barley were grown, particularly in the dry areas of Baluchistan.

- Legumes such as lentils, peas, chickpeas and linum were also major rabi crops. By the late Harappan period, all three were cultivated across the Valley. Pulses include green gram, (moong dal) black gram (urad bean) and horse gram which were grown at several Harappan sites. By 1800 BCE, hyacinth bean and cowpea from Africa were added.

- Rice is native to South and East Asia, particularly long grains in India. In Lothal, Gujarat, and Harappa, rice husks and leaves in Harappan pottery have been found, although it is possible that the rice was a wild crop consumed by grazing cattle. By the Late Harappan Period, wild and cultivated *indica* rice were identified at the site of Hulas. *Japonica* rice was the principal crop of people living at Pirak in the Kacchi plain, Baluchistan.
- Several species of millet were grown in Harappa around 3000 BCE, including little millet, browntop, Job's Tears, sawa, kodon and broomcorn which was found at Pirak in early 2000 BCE. Other African origin crops such as jowar (sorghum,) bajira (pearl millet) and ragi (finger millet) were added to the crops cultivated at Harappa.
- Vegetables like brassica, known as brown mustard, gourds like ivy, cucumber, eggplant and okra were cultivated.
- Jujube, melons, capers, mangoes, sugarcane and grapes were some of the fruits grown. Grapes were cultivated by 3000 BCE and used for grape seed oil. Sesame was the principal plant known for its oil, followed by castor and linseed. Date palms were grown at a very early date in Baluchistan and transported by trade to the Indus Valley.
- Nuts like pistachios, almonds and walnuts were grown or imported, as were garlic, turmeric, ginger, cinnamon and coriander.

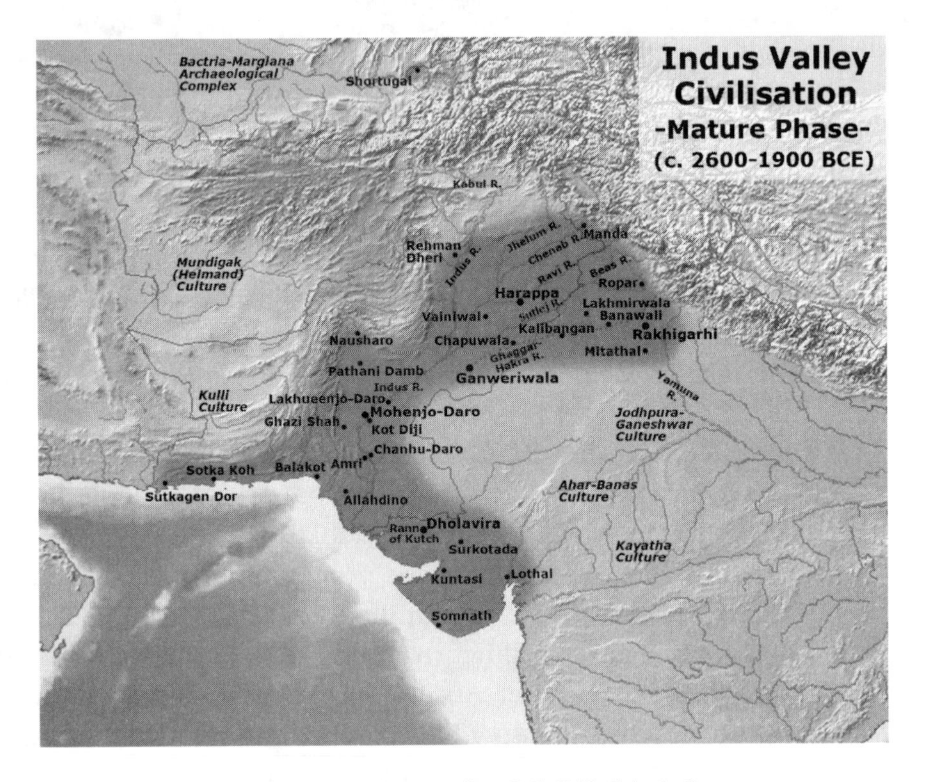

Credit Avantiputra7- CC BY-SA 3.0

Irrigation was necessary in desert like areas such as Baluchistan where small dams such as bunds retained some of the water which flowed in seasonal streams. They also built karizes, which were based upon the Persian qanat system.

- In the Kachi Plain, rivers were used for agricultural purposes, with small dams and channels created on the edge of the plain to retain and distribute the water.
- Shortugai used canal irrigation which they learned from the Namazga culture of Turkmenia.
- The Indus Valley was provided with ample rivers and flood waters for irrigation. Floods which came in July or August provided water for Kharif crops which were sown in June and harvested in October. As in Egypt, the floods brought fine quality silt which made the ground very fertile.

- In Sindh, dams and channels were built to bring water to or carry excess water from marshy areas.
- Harappan farming settlements in Gujarat were confined to regions where rivers and lakes were available along with the Nal depression, which had floodwater in winter. In Kutch, groundwater was used for irrigation.

Tools in the Indus Valley were as sophisticated as those in contemporary Sumer. Lifting devices like shadufs raised irrigation water from wells, streams and channels. They consisted of a long pole with a bucket on one end and a counterweight on the other side. Wooden ploughs with a pointed edge to cut through the ground surface and a curved shaft to draw it along probably had two oxen to prepare the ground. Harvesting was done by flint-edged sickles fastened to a wooden handle.

Food was mainly based upon agrarian cereal and legume produce, along with buffaloes, goats and sheep which provided milk and meat. We know that they consumed meat as it was sometimes included in offerings made for the dead, and for the large quantities of cooked bones discovered. Copper fish hooks, arrow heads and knives indicate that the Harappans hunted, fished and reared animals for consumption. Many fish were dried or salted and eaten at inland cities like Harappa. They also consumed poultry, crocodile and turtle flesh.

Excess grain was stored in large mud-brick granaries. Bread was a very labour intensive task with grindstones and saddle querns which consisted of a flat stone bed with a rounded stone rolled against it, crushing the grain. More *rotis* were eaten in the northern regions, while a liquid porridge consisting of sorghum and millet was being consumed in Gujarat.

The question of rice consumption was a contentious topic amongst archeologists, many of whom believed rice came from China during later times. However, Dr Jennifer Bates of the University of Cambridge, found evidence for a separate domestication process of the wild species *Oryza nivara*. This led to a mixture of 'wetland' and 'dryland' agriculture of *Oryza satvira indica* rice before the true 'wetland' *Oriza sativa japonica* was introduced from China around 2000 BCE. The seasonal multi-cropping strategies of growing summer (rice, millet, beans) and winter (wheat, barley and pulses) requiring different watering regimes preceded ancient Egyptian or Shang Dynasty Chinese agriculture.

REFERENCE: Bhutia, Lhendup, "What They Ate in the Indus Valley," Open Magazine, June 8, 2017

https://openthemagazine.com/features/archaeology/what-they-ate-in-the-indus-valley/

Pottery vessels were used for cooking in a brick-built fireplace. Some ovens resemble modern tandoors. Copper vessels like plates have been found in wealthier homes.

Archeologists Steven Alec Webber and Arunima Kashyap dedicated themselves to studying Harappan cuisine, and made several discoveries. After finding residues of ginger and turmeric on pottery, they determined that the Harappans were making proto-curry. They also discovered that eggplant had been cooked. (See recipe.) Other archeologists such as Dr. Akshyeta Suryanaraya from the University of Cambridge studied lipid (fat) residues in pottery for her doctorate and found an abundance of animal products in the cuisine, such as the meat of pigs, buffalo, sheep, goats and dairy products.

RECIPE- Original Proto-Curry enhanced with Harappan ingredients.

6-7 eggplants slit

1 piece of ground ginger (about 1 inch)

Fresh turmeric ground

1 tbs mango cut into pieces

¼ tsp cumin and salt to taste

Sugercane juice to taste

2-3 tbsp sesame oil

Add eggplants and salt. Cover until nearly cooked through, add water if needed. Stir in mango and cane juice, simmer for a few minutes. Serve with a bajra roti (pearl millet flatbread.)

REFERENCE": Banerjee, Soity, "Cooking the world's oldest known curry," BBC News, June 22, 2016. https://www.bbc.com/news/world-asia-india-36415079

PAKISTAN has been subject to extreme weather disasters since the long drought of 2013. The floods of 2022 were the worst in its history, with as much as one third of the country submerged and nearly half of the crops ruined. The flooding occurred on six fronts:

- Warm weather melted glaciers which sent a torrent of water into several large rivers such as the Indus.
- Karachi was inundated by heavy monsoonal rains.
- Heavy rains in Balochistan washed away homes, farms and topsoil.
- Sindh province experienced flooding from monsoons coming from India.
- Further flash flooding in Balochistan and at the Afghan border

Prior to this catastrophic flood, at least 1.3 million people were experiencing Crisis and Emergency levels of food insecurity in Balochistan and Sindh from the extreme drought.

The impact of the floods on food production in Pakistan has yet to be fully assessed, but the outlook is grim. In 2023, a further 9 million people were propelled into the abyss of poverty, adding to the 153 million people subsisting on an estimated 0.73 US cents per day, according to 'The International News."

https://www.thenews.com.pk/print/1051214-food-insecurity-in-pakistan

VEDIC INDIA

The Vedic age followed the fall of the Harappan civilisation, straddling the Bronze and Iron ages from 1500-500 BCE. Groups of Indo-Aryan peoples migrated into the Indus Valley after 1900 BC and eventually spread their culture and religion into the Ganges Valley. Their literature formed the Vedas, with the *Rig veda* the oldest dating from 1200-1000 BCE, followed by the *Samaveda, Yajurveda and Atharaveda.* By 1000 BCE, they transitioned from semi-nomadic life, where the horse was a revered animal, to settled agriculture in North West India.

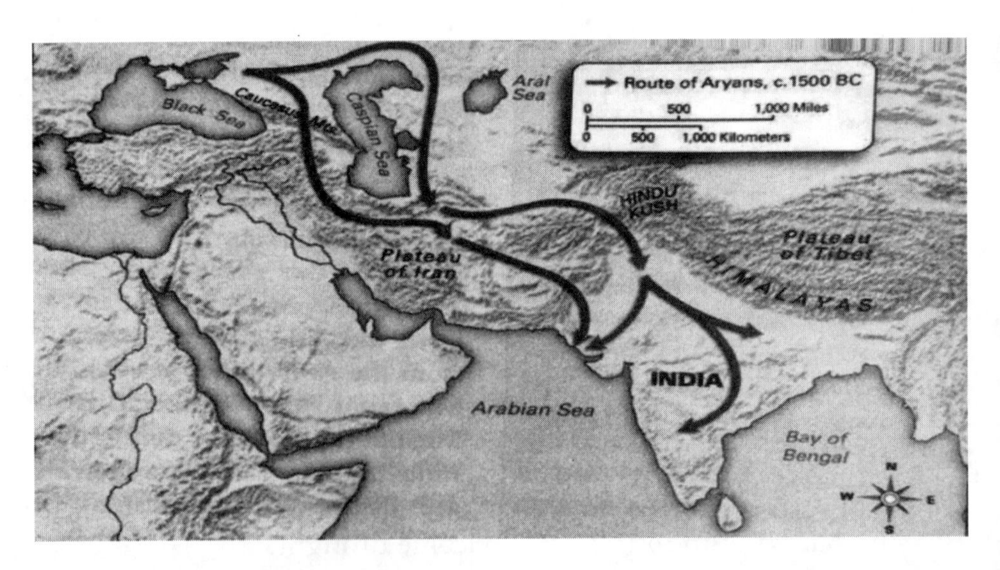

Iron axes and ploughs allowed the inhabitants to clear the thick jungles. Between 1200-600 BCE, Vedic texts including the early Upanishads and Sutras, were completed, and agriculture greatly expanded. During this time the varna caste system was

born, with Brahmins, the priestly caste at the top, followed by the *khsatriyas, vaishyas, shudras* and finally the *dalits, or untouchables* at the very bottom.

Agriculture and pastoralism were the backbone of the Vedic economy. The *Rig veda* referred to the levelling of fields, seed processing and grains stored in large jars. With the introduction of iron implements, crops of wheat, rice and barley were cultivated and stored in surplus. The Vedas describe different cereal grains and their use in daily life. The cow became a sacred animal around 800 BCE, and remains sacred to Hindus.

Food and beverages in the Vedic age included both vegetarian as well as non-vegetarian items. Barley became the staple food of the Aryans, mentioned in the *Rig veda.* Later texts discussed wheat, lentils (red, green and black,) millet and sugarcane.

Cow's milk was a principle ingredient which was consumed after boiling. *Soma,* an auspicious hallucinogenic beverage probably mixed with grains was used in rituals and on special occasions. Buffalo milk, goat milk and cream were also common beverages. Intoxicating beverages were prepared by fermenting fruit extracts; a particularly sweet one was *Soma Ras*. Surd, a spirituous liquor, was commonly used at marriage ceremonies and in some religious rites. It was prohibited for the Brahmins as was *Parisrut,* another semi-fermented liquor. Wine was also mentioned in the Vedas.

Milk products included clarified butter, (ghee) curdled milk (yoghurt), curd, unmelted butter, fresh cream or butter, mixed curd and milk, milk, curd, honey and butter.

Rice was a staple food during the Vedic age, along with barley. It was either *odana,* cooked with water, or *payasa,* cooked with milk. Boiled rice was the most common food in the *Ramayana.* Other popular foods were fried barley cooked with butter, parched rice fried in butter, rice-cakes and rice mixed with milk. Laja, fried rice was also prepared. Salt and sugarcane juice were added to the food as well as pepper and turmeric.

Fruit and vegetables were mentioned in the Vedas. Mangoes, gooseberries and lotus stalks were major vegetables, while cucumbers, olives, gourds and other edible roots were also included.

Dietary restrictions were defined in the Vedic scriptures and based upon the caste system. Brahmins could only eat foods prepared in the finest manner with ghee, (pakka) while everyone else had inferior food (kacca.) Food left on plates was defiled by the eater's saliva and could only be handed to the lowest castes, domestic animals and livestock. Brahmins could not accept food or water across caste lines.

The *Rig Veda* mentions that whatever is offered as a sacrifice to the gods, including animals, was to be consumed by the priest. The *Sutras* and *Aitareya* spoke of the popularity of meat-eating, especially for the king or other highly respected guests. Meat roasted on spits and cooked in pots was an essential element offered to such guests. The *Dharma sutras* defined meat permitted and prohibited, including fowls. Aquatic creatures like the porpoise were also prohibited, along with vegetables such as red garlic, mushroom, leeks and garlic.

AYURVEDA is a traditional Indian medical system which was written down in the early centuries of the Common Era, but existed for up to 1,000 years in the oral tradition.

> "In Ayurveda, regulation of diet is crucial, since it examines the whole human body as the product of food. Ayurveda illustrates how an individual can recuperate by establishing the connection between elements of life, food, and body. According to Ayurveda concepts, food is responsible for different aspects of an individual including physical, temperamental, and mental states. To stay healthy, maintaining a stable healthy diet routinely is essential. The

body absorbs the nutrients as the result of digestion. But Ayurveda states that the food first converts into rasa (plasma), and then followed by successive conversion into blood, muscle, fat, bone marrow, reproductive elements, and body fluids. Imbalance of mind, body, and spirit are referred to as diseases. Ayurveda has different treatments for different diseases, which are well established and consistent over the period."

Sakar, Preetam et al, 'Traditional and ayurvedic foods of Indian origin," Journal of Ethnic Foods, Vol 2, Issue 3, September 2015.

http//www.sciencedirect.com/science/article/pii/S23526181150 00438

There are five elements in the Ayurvedic system: air, fire, water, earth and ether. Each of these combined with another creates a *dosha* a basic energy principle which governs bodily functions. Having too much or too little of these energetic forces causes illness, so the key is balance.

There are multiple therapeutic approaches such as:

- *Shodhana*, a purification treatment of foods, eliminating toxins by peeling, cleansing, sifting, distillation, de-husking etc. For example, cleansing and sifting cumin and mustard seed is known as virecana.
- *Nidan Parivarjan*, avoiding the causative factors for a disease, such as over-eating, uncooked food, sweet food and meat.
- *Shamana* therapy restores the imbalanced doshas by fasting, exercise, sun bathing and digestives.
- *Rasayan*a aims to improve immunity to different diseases. Aswagandha, cucumin, garlic, ginger and neem play an important part.
- *Pathya vyavastha* emphasises diet, activity, and emotional states. A proper diet is essential for good health and healthy body functions. It gives knowledge of foods in

different diseases and helps in the digestion and assimilation of food. (Sakar et el, ibid)

RECIPE: Ghee was considered absolutely pure and used in sacrifice because it was made from cow's milk. This recipe is based upon ancient principles.

Place a pound of butter and place in a pot on the lowest possible temperature, allowing it to simmer. After foaming up, the excess water evaporates and milk solids will come together and eventually precipitate to the bottom of the pot. Pour off the clear fat, which has a deep nutty flavour, and store in a jar.

BUDDHISM, after Hinduism, was the second greatest religion in ancient India. Gautama Buddha (Siddhatta) was born into the aristocratic Shakya clan in the 5th century BCE, but renounced it all to become the Enlightened One who taught how to eliminate sorrow and craving, as well as the cycle of birth and rebirth. After years of asceticism, he discovered the Middle Way, a path of moderation away from the extremes of self-mortification and self-indulgence. Gautama particularly criticised the animal sacrifice rituals as taught in the Vedas, and the Brahmanic superiority in the caste system.

King Asoka was an Emperor of the Maurya Dynasty from 268-232 BCE who promoted the spread of Buddhism across Asia, including Sri Lanka. His rock edicts declare than no animal should be slaughtered for sacrifice, and banned the killing of nursing female goats, sheep and pigs, as well as various types of fish and bulls. He also forbade killing of all fish and the royal hunting of animals. Such edicts protecting animals were almost unheard of in the ancient world.

JAINISM was founded by Parshavantha in the 8th century BCE, and later revived by Mahavira in the 6th century BCE, who preached the doctrine of *ahimsa,* or non-violence to all living things. Their cuisine was and still is completely lacto-vegetarian, and also excludes root and underground vegetables such as onions to prevent injuring insects. Mushrooms and honey were also forbidden as were fermented beverages like wine or beer.

Their very limited diet was composed of green vegetables, fruits, dry beans, lentils, cereals and nuts which contain the least amount of micro-organisms.

FOOD from the Mauryan to Guptan Empires in the 4th century CE showed a shift towards vegetarianism for the Buddhists, Jains and Hindu Brahmans, as well as influences from the Macedonian and later, Roman Empires. Olives, grapes, cumin and coriander were some of the crops inherited from these empires.

Buddhist cuisine was mainly divided into four categories: soft foods like boiled rice, hard food such as roots and fruits, beverages and replenishable articles. Rice was one of the staples of northern India, with at least five varieties mentioned in Buddhist texts. Boiled rice was the favourite, followed by a rice based gruel praised by the Buddhist texts.

Dairy products included milk from cows, buffaloes, goats and camels which was processed into curd. Fresh butter was also found in Buddhist cuisines. Pulses like peas and beans were consumed, particularly in a soup which was made with round fried pulse balls.

Sweets and beverages were prepared with honey or sugarcane juice which was extracted by pressing machines. Sweet rice was made with specific spices and oilseeds. Lord Buddha encouraged his followers to eat such fruits as jojoba, mango and rose apple, as well as rice, barley, vegetables, bread and milk. Meat eating and alcohol were not recommended by the Buddha.

Buddha eating with his monks, Museum of Asian art, Corfu

THE GUPTA EMPIRE was contemporaneous with the late Roman Empire (320-500 CE) and dominated Northern India. Food was discussed in the writings of Kalidasa, Angavijja and the Bharata Samhita, and generally reflected the vegetarian food which was eaten by half the population. They mainly consumed cereals, vegetables, fruits, pulses, breads and milk. Beef was forbidden, but the consumption of goat, sheep and chicken was undertaken by some. The *Puranas* and *Samhitas* describe the medical uses for various food items.

Various rice species were grown, including Sali in Bengal, Sastika and Mahavrihi in Magadha which was used in rituals. Red variety of salt rice was the best for medicinal purposes. Rice was also cooked with curd, milk, ghee, butter, molasses and pulses. Although cow's milk was superior, they also drank the milk of sheep, mares, buffaloes, elephants and camels.

Honey was used extensively in desserts, and more sugar was consumed when the scientists_invented a way to make sugarcane juice into sugar cubes. Spices included ginger, cumin, mustard, coriander, long and black pepper, clove, cardamom and turmeric, while asafoetida was imported from Afghanistan.

Central Asian fruits and vegetables like lemons and carrots reached India in the late Gupta period. Other fruits grown were Kashmir pears, plums, peaches, grapes, pomegranates and melons.

Some intoxicating beverages were prepared from grapes, sugarcane, honey and rice. Various wines were flavoured with mango or patala juice. Upper class women were also encouraged to drink these beverages.

Agriculture during Vedic times involved ploughing with iron implements and sequential cropping was recommended. Animal husbandry was important for dairy, meat, manure fertiliser and for draft purposes such as ploughing. The Mauryan Empire saw the construction of dams and irrigation channels. Horse-drawn chariots were replacing traditional bullock carts.

Greek diplomat Magasthens in his book "*Indika*" provides an eyewitness account of Indian agriculture in 300 BCE.

> "India has many huge mountains which abound in fruit-trees of every kind, and many vast plains of great fertility. The greater part of the soil is under irrigation, and consequently bears two crops in the course of the year. In addition to cereals, there grows millet, and different sorts of pulse and rice throughout India. Since there are two

monsoons in the course of each year the inhabitants gather in two harvests annually."

In southern India, the Tamils were cultivating a wide range of crops such as rice, coconuts, cotton, sugarcane, grains, millets, black pepper, sandalwood, tamarind palms and plantain trees. Water storage systems were built such as the Kallani dam, which is still in use nearly two millennia later.

Sugar was originally refined in the 3rd century BCE according to the Arthashastra, and first crystallised during the Gupta Empire. It was introduced to China via Buddhist monks and by the 6th century CE, sugar cultivation and refining reached Persia, and eventually into the Mediterranean by the Arab expansion.

Early refining methods involved grinding or pounding the cane to extract the juice, then boiling down the juice or drying it in the sun to yield solids which looked like gravel. The Sanskrit word for sugar (Sharkara) also means gravel or sand.

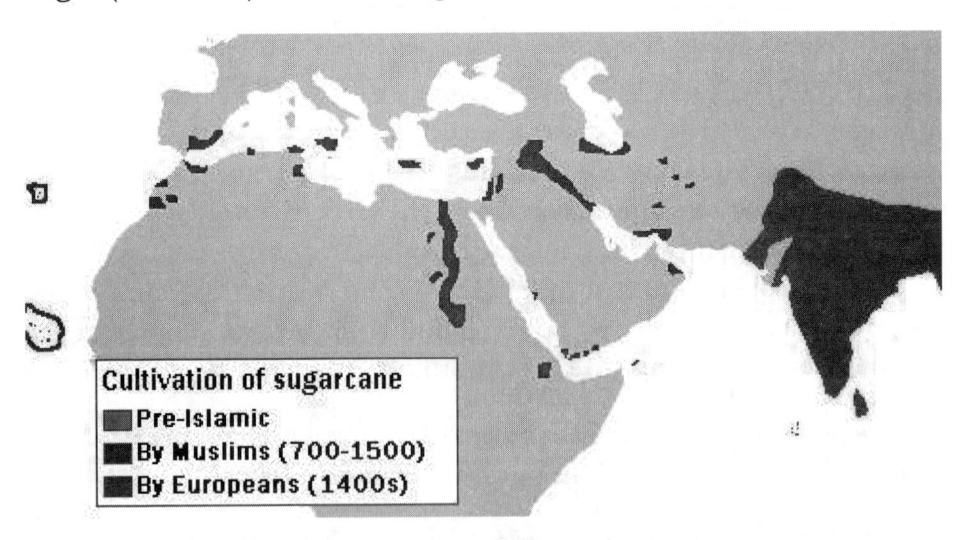

Credit Bless sin

INDIA is ranked at 68 out of 113 major countries in terms of food security index 2022, according to Wikipedia. Food security has always been a major concern in India with nearly 195 million undernourished people according to the UN. Accessing food is difficult for the most impoverished which is why millions of children are malnourished.

India also has a problem with quality protein intake and struggles to make available protein rich foods such as meat, eggs, dairy, lentils and soybeans at affordable prices.

In 2013, the Parliament of India enacted the National Food Security Act providing subsidise food grains for over half the population Various states such as Tamil Nadu and Andrah Pradesh have also enacted initiatives to feed the poor.

Over 60% of the population depend on agriculture, but accessing food because of financial problems causes millions of malnourished children in India

The Covid pandemic and harsh lockdowns in 2020 and 2021 created a food crisis for many who struggled to access enough food to feed their families. Supply chains were adversely affected, and public transport was shut down, stranding millions of migrant workers far from their homes.

The heatwave of 2022 severely reduced the wheat harvest and the country was forced to prohibit wheat exports.

In July 2023, the government of India prohibited the export of non-basmati white rice to other countries to ensure availability in the domestic market. This followed the imposition of a 20% export duty of non-basmatic white rice last September. As India exports nearly 40% of the global rice, this ban will cause considerable increase in world prices and fears of global food shortages.

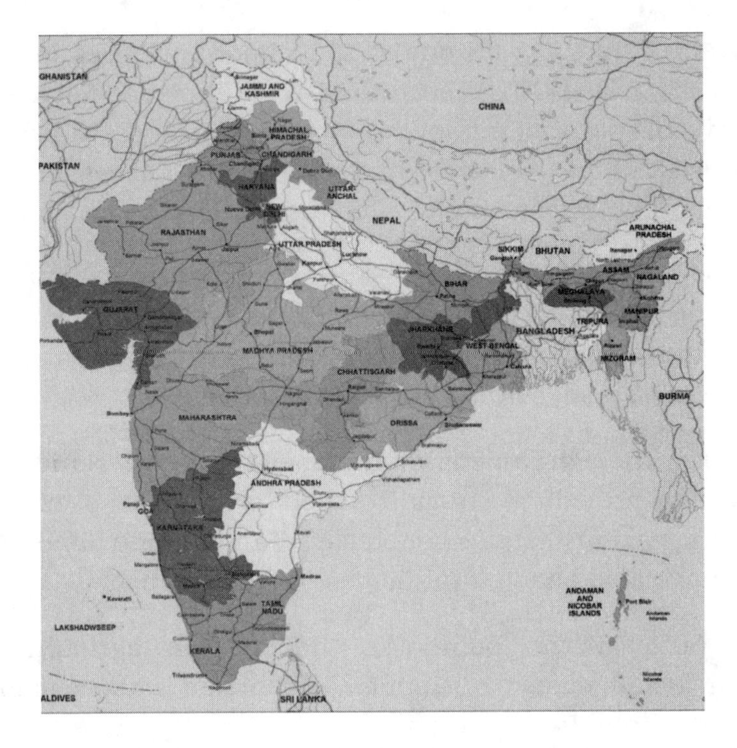

SRI LANKA is known as the birthplace of cinnamon. Its main crop now, as in ancient times, is rice, and huge irrigation projects consisting of a vast network of reservoirs called tanks, allowed the island to grow it in huge quantities over 2,300 years ago.

Agriculture began in the river valleys and the northern and the south-eastern plains, where rice was grown in paddy fields watered by the rains. However, in the dry zone irrigation technology was developed, with the construction of thousands of small irrigation tanks of varying sizes and shapes supplying the water. Rice paddies were cultivated only in the wet season and left fallow during the dry.

Rice cultivation has been dated from 1000 to 300 BCE, with the earliest documented irrigation dating from 300 BCE in the reign of King Pandukabhaya. It was under continuous development for the next thousand years, including underground tunnels, canals, bridges and sluices which were financed by kings. The Anuradhapura Complex was built under King

Dathusena, (459-477 CE) including the massive Kala Wewa reservoir and the 54 (87) Yoda Ela aqueduct to carry the water to Tissa Wewa, with a gradient of only 6-12 inches per mile (10-20 cm per kilometre.)

In the 3rd century BCE, the Sinhalese engineers invented the valve pit which could easily regulate the outflow of water from very large reservoirs. They also invented the small scale Tank Cascade systems in the dry hilly zones, whereby numerous small tanks stored and conveyed water across catchment areas via tunnels and canals. This system was truly as grand as the Roman aqueducts or Mesopotamian irrigation systems.

Credit infolanka.lk

Archeological evidence shows that bananas were originally cultivated in New Guinea 6,800 years ago and were transported to Sri Lanka at least 6,000 years ago, according to phytoliths discovered in Fahien Cave. Because domestic bananas are sterile, mariners must have carried cuttings or whole banana plants on the long journey from New Guinea to Sri Lanka.

The ancient Sinhalese diet consisted mainly of rice with fish, meat and condiments added such as ginger, cardamom, pepper and butter. Animals hunted included deer, elk, buffaloes, wild pig, monkeys, pea-fowl, pigeon and even porcupines. Some tanks had fish which were caught and eaten, but other tanks and reservoirs prohibited fishing. Ocean fish were quite common such as eels and red fish.

They also ate grains, sweet and sour milk as well as butter. Porridge made out of cereal grass was a popular breakfast concoction. Other foods included:

- Cucumbers, melons and bread-fruit were some of the vegetables.
- Pulses including beans and peas as well as root vegetables such as yams and lotus roots
- Fruits mentioned in ancient literature included mangoes, bananas and jack fruit.
- Honey, treacle, sugar juice and raw sugar were added liberally to sweet foods which were usually rice cakes.
- Sweetmeats were made of rice flour.

Coconuts were important ingredients in cooking, especially curries. Coconut groves were planted at least 2,000 years ago. They were also used for beverages, along with milk, sugar cane juice and fruit juice. Fermented coconut milk was the basis of alcoholic beverages.

The Sinhalese ate three meals daily, heela, davala, and rathriya with betel nut chewed between meals. The principal meal, rice and curry, was served at noon. Whilst the king ate off a plate, everyone else ate with their fingers off plantain leaves sitting on the floor. Only the right hand was used for eating.

The earliest mention of food is in the *Arthasastra of Kautiliya* which gives rations of rice, oil and salt that people should eat, such as Aryan and non-Aryan males, women, children, soldiers and kings. In the *Mahavamsa* Chapter XIV, verse 55, King

Pandukabaya feasted honoured guests with "rice soup and with foods-hard and soft." Upon his deathbed, it was read out to another king how he had rice food prepared with honey, lumps of rice with oil and jala cakes baked in butter, but his favourite was Sour Millet gruel.

REFERENCE; "Cuisine of ancient Sri Lankans,"

https://sirimunasiha.wordpress.com/about/food-of-ancient-sri-lanka/

Pepper was added to curries at least 2,000 years ago, along with cumin, turmeric and mustard. Cinnamon, *Cinnamorum verum* is native to Sri Lanka, southern India, Bangladesh and Myanmar. Another species was grown in China and ended up in ancient Egypt and Greece as a very valuable spice.

SRI LANKA was facing multiple crises in 2022, particularly food insecurity. The UN World Food Program (WFP) in June 2022 declared that two in five Sri Lankan households were not consuming enough food.

The serious problem began in 2021 when the Rajapaksa government immediately ordered the country's two million farmers to go organic, imposing a nationwide ban on the importation and use of synthetic fertilisers and pesticides.

The result of this rapid transition was brutal, as domestic rice production fell 20% in the first six months and Sri Lanka was forced to import $450 million worth of rice. At the same time, prices for this staple rose by 50%. The ban also devastated the country's tea crop which was its primary export and source of foreign exchange.

Furthermore, the pandemic destroyed the tourist industry, depriving the country of foreign currency reserves, In November 2021, the government partially lifted its fertiliser ban on important export crops like tea and coconut and offered $200 million to the farmers but it was hardly enough to compensate them for the losses.

The Ukraine war in 2022 was the final straw for Sri Lanka which was unable to access fertiliser. By June the economy, government and currency collapsed and the Prime Minister fled the country.

The EFP sent an emergency relief fund of $60 million to help 3 million vulnerable women and children. The U.S. committed an additional $20 million in aid to feed more than 800,000 children and 27,000 pregnant women over the next year.

Inflation of food was at a record 90%, making even rice unaffordable to millions of families. Soaring fuel prices have also had a disastrous effect on the economy, with around 200,000 fishermen unable to work their boats because of fuel import restrictions. By mid-2023 conditions were beginning to improve as tourism returned and farming restrictions were removed.

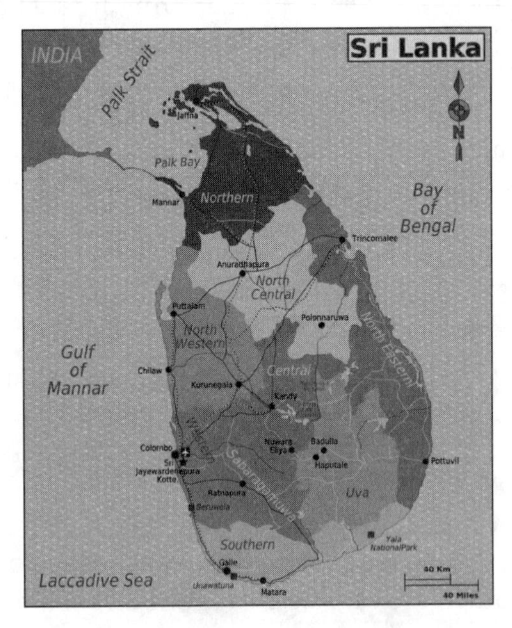

INDONESIA is a vast archipelago of thousands of islands, including Java and Bali in South East Asia, straddled between the Indian and Pacific Oceans.

The **Moluccan Islands** were known to early Asian and European explorers and traders as the Spice Islands, because crops such as nutmeg and cloves were first cultivated on the island of Banda. Agriculture in the Indonesian islands developed millennia ago, with rice, coconut, sugar palm, tubers, sago, taro and tropical fruits among the earliest produce cultivated. Wild rice was cultivated on the island of Sulawesi from 3000 BCE and spread to Java 3,000 years ago when the Dong Song culture brought wet-rice cultivation, ritual buffalo sacrifice and bronze casting from Vietnam.

Rice production was linked to the development of iron tools which helped clear the forest, and the domestication of the Asian water buffalo for rice paddy cultivation and manure. Images on the stone temples of Borobudur and Prambanan in Java show rice cultivation, rice barns, buffalo driven ploughs, women planting seedlings and pounding the grain.

The Hindu goddess Dewi Sri was worshipped in ancient Java and Bali, and rice harvest festivals were celebrated in Sundanese rituals. In Bali, the traditional subak irrigation system was created to ensure adequate water for the rice paddies.

The **Balinese Subak irrigation system,** which was developed in the 9th century CE, was comprised of five terraced rice fields, water temples, canals, tunnels and weirs. The temples had the practical function of regulating the flow of water from springs and canals downhill to the paddy fields and were managed by the priests.

Balinese rice terraces. Credit Imacim, CC BY-SA 4.0

The word *subak* is an old Balinese word refers to a unique social and religious institution which survives today as a democratic association of farmers who determine the use of irrigation water for growing rice. The areas of the current subak system include these areas:

- Supreme Water Temple of Pura Ulun Danu Batur
- Lake Batur
- Subak Landscape of the Pakerisan Watershed
- Subak Landscape of Catur Angga Batukaru
- The Royal Water temple of Pura Taman Ayun

INDONESIAN CUISINE shares similarities with Malaysian food as well as many regional culinary traditions which were based upon indigenous culture mixed with some foreign influences. Food, recipes and spices from China, India and the Middle East had been adopted due to extensive caravan trading routes extending back into prehistoric times. Javanese cuisine is the most authentic with a hint of Chinese influence, while Sumatran food has Indian and Middle Eastern food influences. Rice was introduced as early as 2300 BCE and by 2000 BCE Chinese traders brought cabbage, mustard and soybeans as well as fried rice, which continues today as the famous nasi goreng. Trade with India introduced curries, cucumbers, onions, mangoes and eggplant along with spices by 100 CE.

Ancient dishes were described in many Javanese inscriptions, such as *hadanan Haran*- minced water buffalo satay, *hadaan Madura*- water buffalo meat simmered with sweet palm sugar and *dundu puyengan*- eel seasoned with lemon basil. Grilled meat included pork, water buffalo, deer and goat, while boiled vegetables like yams and tubers were served with liquid palm sugar. Ancient beverages included sugar, jasmine and tamarind juices. The Old Javanese *Kakawin Ramayana* mentioned cooking techniques when Sita offered delicious food cooked with oil.

RECIPE: Rawon is a Javanese beef soup first mentioned as *rarawwan* in a Javanese Taji inscription 901 CE. Its ingredients include a ground mixture of garlic, shallots, keluak nut, ginger, candlenut, turmeric red chilli and salt which are sautéed in oil. The mixture is then poured into boiled beef stock with diced beef slices. Finally, bay leaves, lemongrass, galangal, kaffir lime leaves and sugar are added as seasonings. The black soup, coloured by the keluak nut, is garnished with fried shallot or green onion.

REFERENCES; Wikipedia, Indonesian cuisine, Rawon

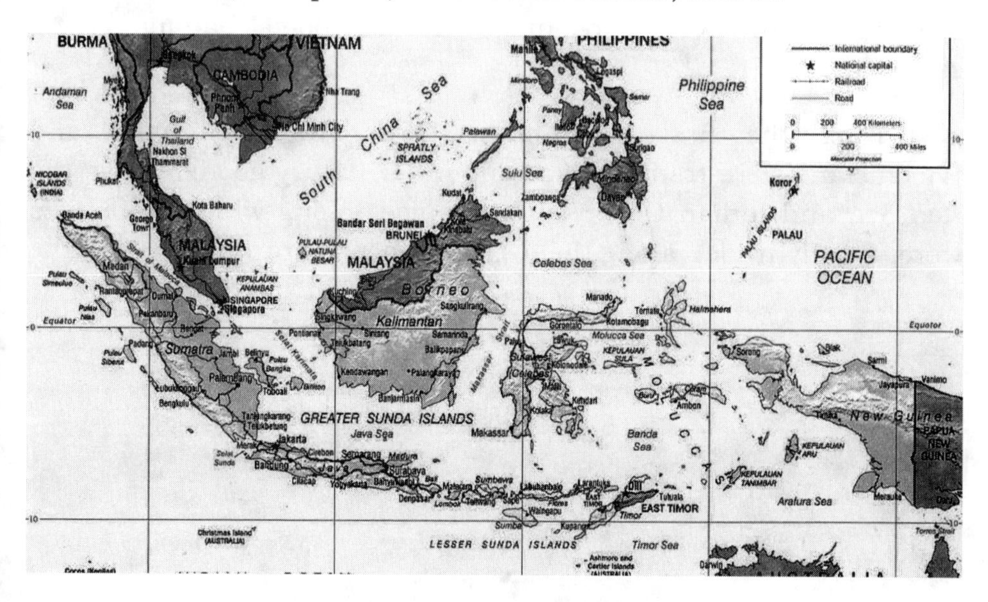

Credit, Indonesian Wikipedia

THE ANCIENT SPICE TRADE routes extend back thousands of years to Neolithic times with spices, obsidian, shells and precious stones traded. The Egyptians were trading with the Land of Punt in the 3rd millennium BCE which was on the Horn of Africa.

A spice is a fruit, seed, bark or other plant substance used for flavouring or colouring food. Over long caravan trading routes, such spices reached Egypt, and cloves which were native to thc Moluccan Islands, were used in Mesopotamia by 1700 BCE. The Egyptian Ebers Papyrus, which dates from 1550 BCE, describes many herbal medicinal remedies and potions containing imported ingredients.

Maritime routes across the Indian Ocean developed before 1500 BCE by the Austronesian peoples of South East Asia who built the first ocean-going ships which established trade routes with India and Sri Lanka. Indonesians were trading in spices, particularly cinnamon, cassia and nutmeg with East Africa using catamarans and outrigger boats. Other foods traded across the Indian Ocean were coconuts, sandalwood, bananas and sugarcane.

During the first millennium BCE, Arabs, Indians and Phoenicians were trading spice and other luxury goods across the Red Sea and Indian Ocean. Indian spices along with fine textiles were usually traded along the overland incense route.

nutmeg and mace

These cultures used spices to make incense, ointments, and to cover the lack of freshness in their foods. In return, the traders brought gem stones, ivory, frankincense and animal skins from the Middle East/African coastal cultures.

The Egyptian port of Alexandria became a major trading centre for spices from the east. In the 1st century CE, sailors learned to sail across the Indian Ocean by reading the rhythms of the monsoonal winds, which opened up quicker and more effective trading routes. The links were formed by traders buying and selling goods from port to port

- Nutmeg or *Myristica fragrans*, native to the Bandas, Indonesia, is the seed of the fruit, while mace comes from the seed covering. The earliest usage of nutmeg comes from the island of Pulau Ai in the Banda Islands where a 3,500 year old potsherd had residues from the plant. In the 6th century CE, nutmeg spread to India and eventually to Constantinople.
- Cinnamon is native to Sri Lanka.
- Black pepper is native to the Malabar Coast of India.
- Cumin's origins are either Central Asia or South West Asia.
- Coriander is native to Southern Europe, Northern Africa and South Western Asia.
- Garlic was first cultivated in Central Asia and spread from Egypt, China to the Mediterranean.
- Basil is native to Central Africa and South East Asia.
- Ginger originated in China.

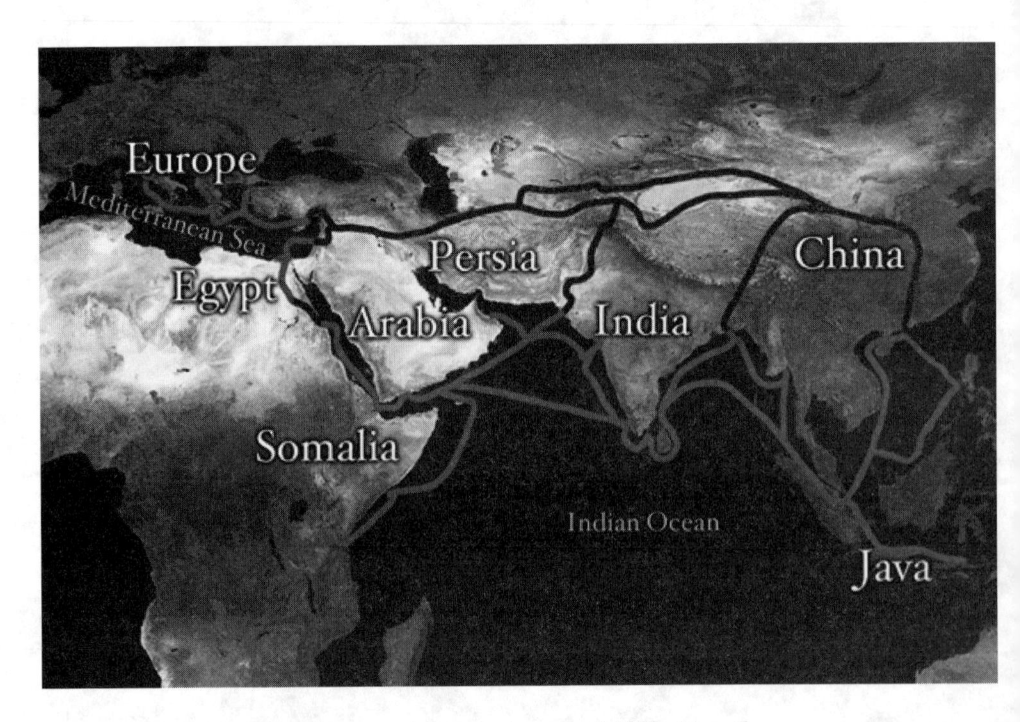

Ancient spice routes and Silk Road

INDOCHINA- Vietnam, Cambodia and Thailand

Belonging to South East Asia, along with Indonesia, these countries were also the earliest to cultivate taro, yam and citrus fruits. Mystery surrounds the origins of agriculture in this area, but the prevailing theory is that rice cultivation was imported from China between 4,000 and 3,000 years ago, based upon limited archeological excavations. In 2018, a study of 4,000-year-old DNA from skeletons in Man Bac, Vietnam, discovered that these early agriculturalists shared physical and cultural similarities with inhabitants from southern China. A later influx of southern Chinese agriculturalists brought bronze metalworking technology 2,000 years ago to the region.

REFERENCE: Wade, Lizzie, "How rice farming may have spread across the ancient world," ""Science," May 17, 2018. https://www.science.org/content/article/how-rice-farming-may-have-spread-across-ancient-world

THAILAND

Spirit Cave in Thailand was occupied from 12,000 to 7,000 years ago by hunter gatherers of the Hoabinhian culture. First excavated in the mid-1960s by Chester Gorman, it was found to contain remnants of almonds, betel, broad bean, pea, bottle gourd, chestnut, butternut, candle nut and pepper in layers dating to around 9800 to 8500 BCE. Although they were no different from their wild phenotypes, Gorman suggested they had been used as food, condiments, stimulants and even some early domesticated plants. Later he wrote:

> "Whether they are definitely early cultigens remains to be established... What is important, and what we can say definitely, is that the remains indicate the early, quite sophisticated use of particular species which are still culturally important in Southeast Asia."

Gorman C. (1971) The Hoabinhian and After: Subsistence Patterns in Southeast Asia during the Late Pleistocene and Early Recent Periods. World Archaeology 2: 311

In 1972 W.G Solheim, director of a Spirit Cave project wrote an article in "Scientific American" noting that while the specimens may have been wild species, the inhabitants had a knowledge of horticulture and ceramic technology. Although ceramics were dated no earlier than 6800 BCE, Solheim theorised that ceramic technology was invented 15,000 years ago, without being able to substantiate this claim. More recent studies indicate that the organic resin on some of the sherds is only around 3,000 years old, and the general consensus today is that the plant remains found in Spirit Cave were collected, but not cultivated by people of the Hoabinhian culture.

Around 1970, pottery decorated with elegant cords was found near the village of Ban Chiang in northeast Thailand dated back 7,000 years. The inhabitants cultivated rice, and after 1700 BCE, used bronze tools such as spearheads, axes, adzes, hook and blades.

By 4700-4000 BP, domesticated rice and shell sickles were found at the Khok Phanom Di site in Thailand. The earliest evidence of millet in Thailand is dated to 2300 BCE. The first Neolithic people at Ban Non Wat grew rice, fished, kept domestic animals like pigs and collected shellfish from 1700 BCE.

Earthenware water buffalo, 2300 BCE, credit PHGCOM CC BY-SA 4.0

Ancient Chinese settled Indochina, including Myanmar and Thailand around 600 CE, bringing stir-fry and deep-fry techniques. The steel wok became an essential cooking item and noodles were widely adopted. The Thais started creating rice noodles, as well as the usual egg and wheat noodles. Indian influences on the cuisine included spices for curries and herbal medicine, using plants like bunnak or rose chestnut. Turmeric, coriander, cumin and other Indian spices were used to make colourful curries, along with indigenous ingredients like lemongrass, Thai holy basil and galangal (Thai ginger.) Other culinary influences from Malaysia like sambal and red curry, known as Panang curry, and Burmese curries were also introduced.

By the 13th century, the Thai cuisine was based around various types of meat and seafood combined with vegetables, herbs and spices such as garlic and pepper served with rice. Chillies did not enter the cuisine until the Portuguese introduced them from South America two centuries later.

Royal Thai cuisine is the apex of Thai food, believed to have been created during the days of the ancient Ayutthaya court (1351-1767.) The Siamese kings kept vast palace kitchens manned by massive brigades, led by foreign and local chefs. Some favourite dishes were *sangkaya fak tong* (steamed pumpkin stuffed with coconut custard,) *mieng kum* (wild betel leaf-wrapped appetiser) and *gaeng ranjuan* (beef soup with fermented shrimp paste.)

VIETNAM

The earliest Neolithic sites have been found along or near present or former coastlines and rivers as they provided the flooding required for rice cultivation. Vietnam has the Mekong River in the south and the Red River delta in the north adjoining China. The settlement of An Song along the Vam Co Dong River overlooked alluvial floodplains with rice fields. The inhabitants were involved with fishing, animal husbandry (pigs) and ceramic production.

The Lạc Việt of the rice-farming Phung Nguyen culture in the Red River basin were predominantly descendants of ancient agricultural communities of the Yangtze and southern and central China region, who arrived in Indochina around 2000 BCE. The fertile soil along with adequate rainfall led to a prolific growth of rice and other plants. During the rainy season, the villages had to manage floods, transplant rice and harvest it. One of the core values of village life was to live in harmony with nature and others.

Fishing and hunting supplemented rice and betel nuts were widely chewed. From 2000 BCE, stone tools became more sophisticated, to be replaced by bronze tools a millennia later, including axes and sickles. After 1000 BCE, the Vietnamese were skilled rice farmers and fishermen who kept buffaloes and pigs.

Vietnam was incorporated into China in the 2nd century BCE and remained a Chinese province for the next thousand years. Noodles made of millet or other native grains were invented in China during the East Han Dynasty from 25-220 CE. During this time, large scale wheat grinding became available, providing flour to make *mi* or *mien*. The Chinese introduced several dishes such as wonton, moon cake and *chow mein*. Ethnic mountain tribes near the China/Vietnam border adopted roasted pork and pork belly from China.

Spices were brought to Vietnam via Malay and Indian traders, including curries made from chicken and goat. Coconut milk was influenced by Cham cuisine. Rice was a staple, as it was throughout Asia.

Vietnamese fish sauce, ***nuoc mam*** may have had its origins in ancient Carthage, as the oldest records of the sauce were found in the writings of Mago. After the Roman conquest of Carthage, fish fermentation techniques were spread across the Roman Empire where it was known as garum. Archeologists believe the methods of making fish sauce spread into Asia along the Silk Road during the late Roman Empire. From the 5th century onwards, for nearly one thousand years, fish sauce was the main condiment in Asia until the Chinese invented soya sauce in the 14th century.

CAMBODIA

Excavations at Laang Spean cave confirmed the presence of Hoabinhian stone tools from 7000 BCE and pottery from 4200 BCE. Rice was cultivated up to six thousand years ago, and of the 2,000 indigenous rice varieties, *malys Angkor* is regarded as the finest.

The first rice farmers were found at Samrong Sen and may have migrated from southern China around 1500 BCE. During the Iron Age, beginning in 500 BCE, the villagers lived along the

Mekong River in houses on stilts, where they cultivated rice, domesticated animals and fished.

The Funan Kingdom (100-627 CE) which incorporated Cambodia and South Vietnam, had urban centres along with surplus food which was traded by sea to various ports. Trade with China and India expanded, as did the Hindu religion. Indian merchants introduced curries and spices around the 2nd century CE. Coconut based curries as well as boiled red and white sweets were all influenced by Indian cuisine.

The Khmer Empire, centred in Cambodia, also covered parts of China, Burma and all of Thailand and Vietnam from 802-1431 CE. Ruled as an absolute monarchy, Hinduism and Buddhism were the state religions.

The Khmers were traditional rice farmers, who built massive hydraulic complexes including giant reservoirs called **barays** and a network of canals. Fish were also a huge part of their diets, with most coming from freshwater Tonle Sap. The Chinese started arriving in the Khmer Empire in the 13th century, bringing noodles and stir frying.

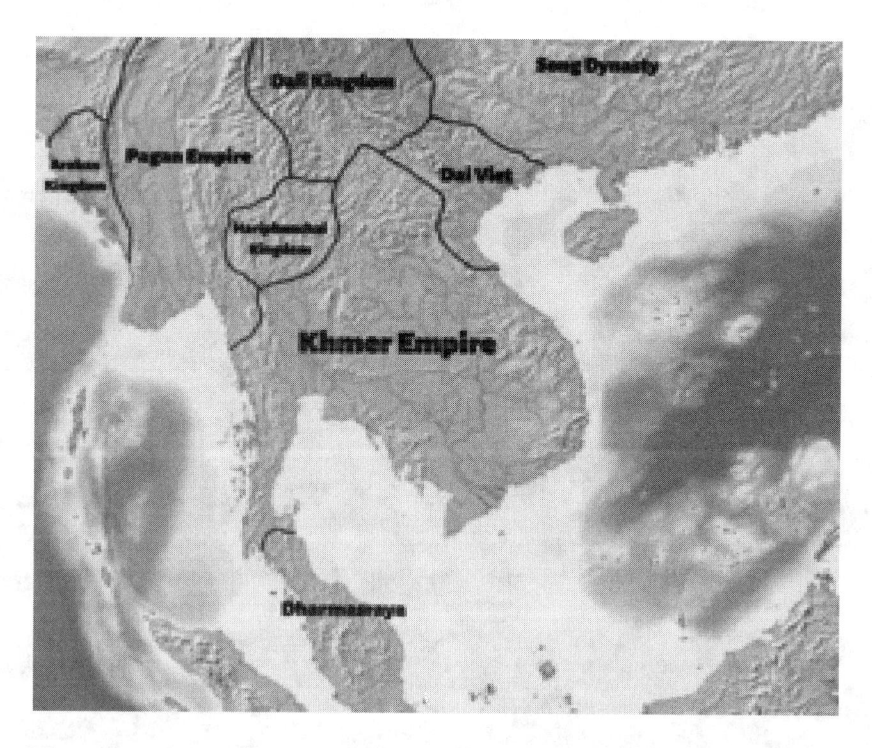

Khmer cuisine was classified as royal, elite and peasant. Many ancient recipes were lost forever during the brutal regime of the Khmer Rouge in the late 20th century, but some were rescued by Cambodian chef Ly San, who interviewed many elderly survivors of the regime. They taught him that broths, curries, roasted meats and dipping sauces had been part of traditional Khmer cuisine.

Royal food included *na tang*, a starter of deep fried sticky rice with a sauce made from minced pork, chilli and coconut milk. Such palace dishes were richer and contained more meat. Elite dishes, for officials and wealthy citizens, were made using lower quality meats and ingredients. Peasants ate *kor kou*, a fish soup with simple ingredients.

The link to the original CNN Travel article at https://tinyurl.com/2p946vaa can no longer be found.

Carving Angkor Wat

Prahok is a Cambodian staple fish paste with a long history going back to 100 BCE. Originally it was created to preserve fish during the dry season. Made from mud carp, it was originally crushed underfoot before being salted and fermented in baskets. The liquid from the fermenting fish is a fish sauce which is they dried and salted again before being placed in containers for up to two years. The f1avour is very pungent.

CITRUS

It is believed that the genus citrus originated in the foothills of the Himalayas, from Assam to northern Myanmar to western Yunnan around 5 million years ago. Wild fruit plants included citrons of South Asia, pomelos of South East Asia, cumquats and mandarins of South Eastern China, and kaffir limes of the South East Asian islands. Later it spread into Taiwan and Japan with the tachibana orange and into Papua New Guinea and northern Australia by 800,000 years ago.

Various species of citrus were native to South East Asia, South and East Asia as well as Melanesia and northern Australia. From 3000-1500 BCE their cultivation spread to Micronesia and Polynesia. Via the incense trade route, the citron was transported to the Middle East by 600 BCE where it was grown in a Persian style garden in Jerusalem. During Hellenistic times, the citron was brought to Europe.

Citron CC BY-SA 3.0

Archeobotanist Dafna Langgut of Tel Aviv University has traced the spread of citrus from South East Asia to the Mediterranean. Using ancient texts, murals, coins as well as fossil pollen, seeds, charcoal and other fruit remains, she traced the citron to the Hellenistic empires by the third and second centuries BCE. The oldest lemon was found in Rome in the first century BCE, where it was favoured by the Roman elite for medicinal purposes and its pleasant odour.

Oranges are a hybrid of pomelo and mandarin which were first cultivated in China. From China they spread to Central Asia and the Arab Empire, where they were widely cultivated in Spain in the 10th century, supported by extensive irrigation networks.

However, citrons were grown in the Middle East much earlier. It wasn't until a millennium later that sour oranges, limes and pomelos were brought by Muslims to southern Europe from northern Africa.

REFERENCE: "A History of Citrus Fruit," "Archeology,"

https://www.archeology.org/news/5758-170724-archeobotany-citrus-fruit

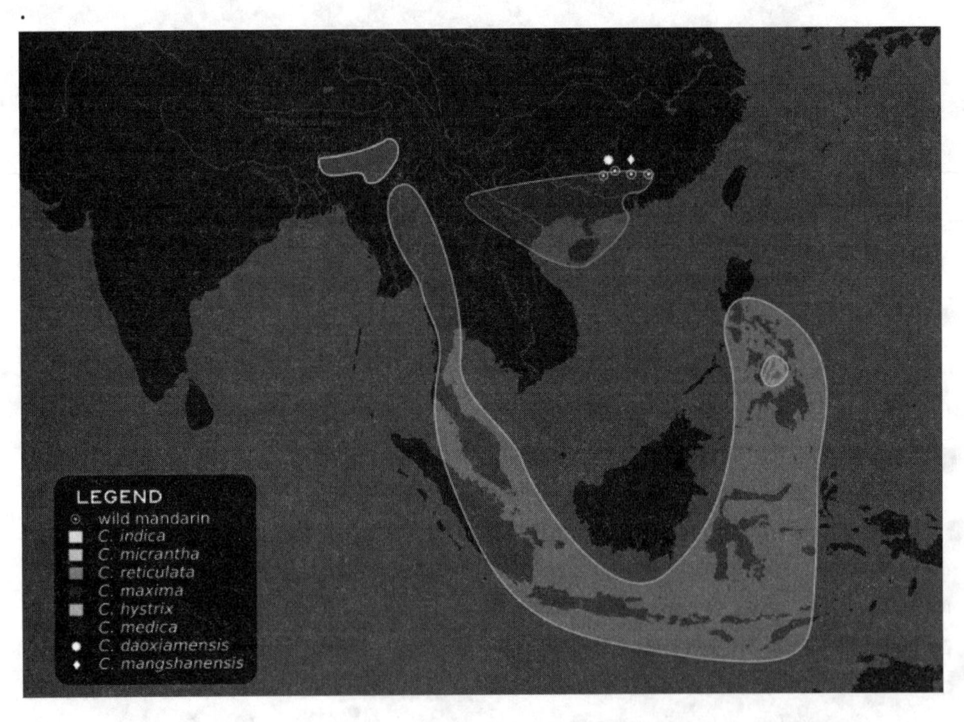

Ancient wild citrus homelands. Credit Obsidian Soul CC0

Pomelo, credit Ivar Leidus CC BY-SA 4.0

CHINA was one of the earliest sites for agriculture in the world. It is also the longest continuous literate civilisation.

The **Neolithic** period began around 10,000 BCE and concluded with the introduction of metallurgy 8,000 years later. These Neolithic civilisations grew up around two major river systems, the Yellow (central and northern China) and the Yangtze (southern and eastern China.) They were characterised by settled communities relying on farming and domesticated animals rather than hunting and gathering.

The **Pengtoushan** culture centred around the Central Yangtze River region in Hunan province is dated between 7500-6100 BCE.

Rice residues have been carbon dated to 8200-7800 BCE, showing that rice was already domesticated at this time. Domesticated rice grains were larger than naturally occurring rice. Pottery was in use by 5800 BCE.

The Yangshao culture along Yellow River in the Gansu/Henan provinces in North Western China was known for its ceramics around 6,000 years ago. This clay pottery was fired at about 500-600 degrees C and decorated with colours, human, animal and geometric designs. Silk production, known as sericulture also began on a small scale, as did woven hemp for clothes.

The main Yangshao crops were millet, both foxtail and prosco millet, with some early evidence of rice cultivation. Middle Yangshao settlements such as Jiangzhi had raised floor buildings for grain storage as well as grinding stones for making flour. They kept pigs, sheep, goats and dogs for meat, and used stone tools for hunting and fishing.

The **Longshan** or Black Pottery Culture was a late Neolithic culture in the lower Yellow River valley of northern China from 3000-1900 BCE. Pottery making, with the use of pottery wheels, reached a high standard. The most important crop was foxtail millet, along with broomcorn millet; rice and wheat were also grown. Specialised tools for digging, harvesting and grinding grain were recovered from sites. Pork was the most common source of meat, while sheep and goats were domesticated 6,000 years ago in the Loess plateau.

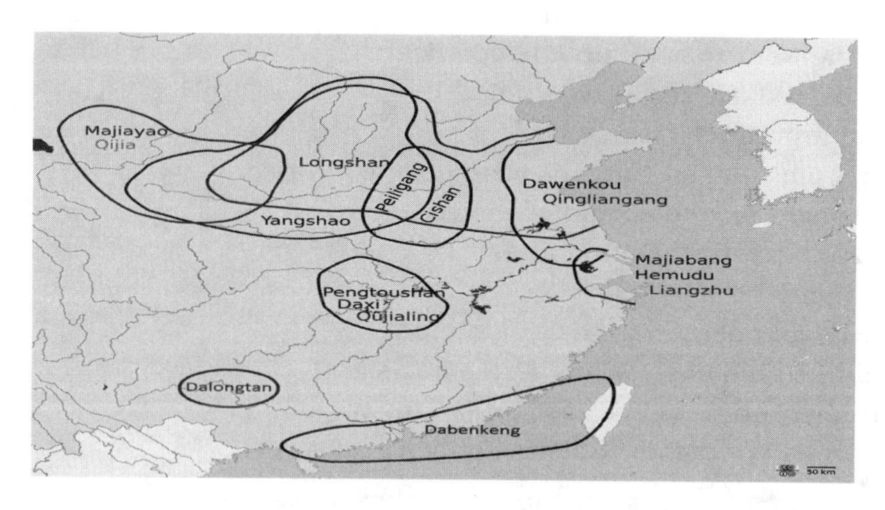

Neolithic map of China

Chinese agricultural myths were codified and transmitted by the Han Chinese from much older myths. The Emperor god Shennong is credited with inventing agriculture, including the plough, axe, hoe, irrigation, digging wells and herbal medicine. He was also an expert on plants and poisons, claiming the "five sacred grains" to be millet, wheat, soy, rice and beans. The *□□□11□1□□□□1□ăo J□ng* book on agriculture and medicinal plants, attributed to Shennong was written down around 250 CE as a compilation of older oral traditions.

An oral epic poem, titled the *Hei'anzhuan* ("Story of Chaos") describes how Shennong found the seeds of the Five Grains:

Shennong climbed onto Mount Yangtou,
He looked carefully, he examined carefully,
Then he found a seed of millet.
He left it with the Chinese date tree,
And he went to open up a wasteland.
He planted the seed eight times,
Then it produced fruit.
And from then on humans were able to eat millet.
He sought for the rice seed on Mount Daliang,
The seed was hiding in grasses.

He left it with the willow tree,
And he went to open up a paddy field.
He planted the seed seven times,
Then it produced fruit.
And from then on humans were able to eat rice.
He sought for the adzuki bean seed,
And left it with the plum tree.
He planted it one time.
The adzuki bean was so easy to plant
and was able to grow in infertile fields.
The soybean was produced on Mount Weishi,
So it was difficult for Shennong to get its seeds.
He left one seed of it with a peach tree,
He planted it five times,
Then it produced fruit,
And later tofu was able to be made south of the Huai River.
Barley and wheat were produced on Mount Zhushi,
Shennong was pleased that he got two seeds of them.
He left them with a peach tree,
And he planted them twelve times,
Then later people were able to eat pastry food.
He sought the sesame seed on Mount Wuzhi,
He left the seed with brambles.
He planted it one time.
Then later people were able to fry dishes in sesame oil.
Shennong planted the five grains and they all survived,
Because they were helped by the six species of trees.

Other venerated agricultural deities include Houji, Houtu, an Earth deity, Shujun, a god of farming and cultivation; also known as Yijun and Shangjun. He was credited with inventing the use of draft animals to pull ploughs to turn the soil prior to planting. Suiren, also known as “Drill man” used a fire-drill to allow food to be cooked. He was also involved in teaching animal husbandry, the invention of various tools, the domestication of rice and ginger as well as irrigation by digging wells. Other myths include the beginning of farmers’ markets and the invention of fermentation.

Huangdi, the Yellow Emperor from 2699-2588 BCE, was also credited in ancient texts as the first teacher of cultivation.

Han Dynasty painting of Shennong

BRONZE AND IRON AGE AGRICULTURE

The **Shang Dynasty** ruled from 1750-1085 BCE and used bronze metallurgy to make weapons which allowed them to take over the previous semi mythical Xia dynasty. The dynasty was situated around the Yellow River where the fertile sediment called loess was deposited, allowing for the development of agriculture.

The Shang economy was feudal, with the Emperor owning all the lands which were worked by peasants who could only keep a small amount of their produce. The peasants were also conscripted to work on building irrigation canals or serve in the army. Farmers were using simple tools or stone and wood, even though bronze metallurgy was developed. Main crops were wheat, millet, rice and barley, while fruit, vegetables and nuts were also grown.

The Shang Dynasty was replaced by the **Zhou** (1046-226 BCE) with the dynasty beginning to dissolve in its final years in a phase known as the Warring States Period (476-221 BCE.) The King owned all the land and the farms were known as "salary settlements" which were not able to be subdivided, inherited or sold. Only the king had the right to confiscate it and allocate it to other functionaries. The **well-field system** was a 9-field and allotment method of taxation which began in the Zhou or late Shang era. The name is derived from irrigation and drainage canals built in the fields and the demarcations of the outer border. The grain of the middle square was taken by the government, while individual farmers could keep grain cultivated in the other squares.

Iron metallurgy was developed in the Zhou and was well established by the **Spring and Autumn** period around 600 BCE for weaponry and farming tools, particularly the iron plough pulled by oxen or buffaloes. Eventually oxen were yoked in teams, and previously untillable land became fertile. A device known as a well sweep, a bucket suspended at the end of a long pivoting lever greatly enhanced small scale irrigation.

The farmers used a grid system, irrigation, fertilisation and crop rotation as well as iron ploughs drawn by oxen from 400 BCE. Huge hydraulic projects were begun in this dynasty for irrigation and flood mitigation such as the Dujiangyan Irrigation System built by magistrate Li Bing in 256 BCE.

Irrigation improved with the invention of the chain pump in **Han** times which brought water from low areas to the fields. This device consisted of a continuous moving chain with plates that scooped up water on one side of the pump to drop on the other side. The machine was operated with a treadmill. Other Han inventions included:

- The wheelbarrow which consisted of a cart attached to a single wheel that helped carry crops with more efficiency.

- The waterwheel had buckets attached to scoop up water from rivers or lakes.
- Multi-tube iron seed drills allowed seeds to be positioned in the soil and buried at the correct depth.
- The Wind winnower separated grain from the seed casing. The Han farmers used a rotary winnowing fan which used a cranked fan to produce the airstream.

Chain pump

The "*Treatise on Food and Money*" by Ban Gui provided valuable information on agriculture during the Han Dynasty. Excerpts include:

> "In sowing crops, there was always a mix of the five major grains in order to guard against calamities (blights affecting one crop). It was not permitted to plant trees in cultivated fields as they would hinder the growth of grains. Ploughing was done with energy and the fields were frequently weeded; the harvest was reaped as though bandits were about to appear. Encircling the cottages, mulberry trees were planted.

"Vegetables were planted in garden plots, and at the borders of living and working areas were planted melons and gourds, fruit trees and cucumber. Chickens, pigs, dogs, and swine were raised for food with close attention to their timely needs. Women tended silkworms and wove the silken cloth. In this way, persons above the age of fifty could be clothed in silk, and those above seventy always had meat to eat."

Pickled cabbage, known as *suan cai,* was first mentioned in the "Book of Odes," dating back to the 7th century BCE. Fermented vegetables were used as offerings in the ceremony of worshipping the ancestors. Centuries later, the workers building the Great Wall of China lived on rice and cabbages. To preserve the vegetables in winter, they added rice wine to the cabbages, which fermented them with a sour taste.

Suan cai is still eaten in China, with variations in Vietnam and Thailand. Northern Chinese used napa cabbage, while Chinese mustard has long been fermented in southern and western China. From Asia, fermented vegetables spread to eastern and central Europe via invasions by the Mongols. By the 16th century, it became known as sauerkraut in Germany, created by the production of lactobacillus bacteria in the fermentation process.

The Tang Dynasty from 618-907 CE saw many changes in agriculture with the widespread cultivation of garlic, soybeans and peaches. Tea, which had been popular since the 5th century, became the beverage of choice for most people.

CROPS

Rice farming in the southern Yangtze River basin, China began in the Neolithic, around 9,000 years ago, with the cultivation of *Oryza satvia japonica.* According to legend, Shennong was the first to initiate the planting of rice. The wet rice method, using a paddy field, requires water flooding from the rain, river or irrigation and a relatively warm climate. The water creates a stable environment for the rice to grow without extremes of temperature and the burden of weeds. Terrace farming allows the water to fall from the higher level paddies to the lower terraces.

Harvesting rice involves draining the field, waiting for the rice to dry and then cutting it and gathering into sheaths. When the grain is separated from the stalks and dried, husks are winnowed, a process which separates the husks from the chaff.

Rice terraces, Yunnan. Credit Jialiang Gao, CC BY-SA 3.0

During the Zhou Dynasty, rice was popular throughout all levels of society. Ploughing would take place in the spring, weeding in the summer, harvesting in the autumn and storing in the winter. Rice was also used to brew wines and offered as a sacrifice to the gods. Books discussing rice agriculture appeared during the Warring States Period. By the Han Dynasty, rice was eaten by all segments of the population.

Rice wine had been brewed before the Shang Dynasty, and inscriptions on the bones from the 16th century BCE mentioned three types of alcoholic drinks made from cereals and wine. Jiu Qu, rice wine, was popular with the Zhou court and highly regulated. During the Han, production expanded, and by 600 CE a book on fermentation, including rice wines, was published by Jia Shixi.

Soybeans originate from China. Emperor Shennong named soybeans as one of the five sacred plants. Charred remains of wild soybean *Glycine soja*, were discovered in Jiahu, Henan province, a Neolithic site occupied between 9,000 and 8,000 years ago. They were domesticated between 1800-1100 BCE in eastern China during the Shang Dynasty and cooked or fermented into a paste. The oldest preserved soybeans resembling modern varieties were found in Korea and dated to 1000 BCE.

Soybeans were an important crop during the Zhou Dynasty but unknown in South China before the Han period. The production of **tofu**, curdled soy milk or bean curd, allegedly occurred during the Han period, although its origins are shrouded in legend. One fable claims it was masterminded by King of Huai-nam, Liu An, while another legend credits its discovery with a man who created tofu as an easy way to feed his elderly parents. Or it could have been invented by copying the cheese-making methods of Mongolian herders who drank milk.

Soy sauce began to appear more than 2,500 years ago when soybeans were fermented with wheat and salt which was expensive. It was boiled in water, cooled down to 80 degrees F and inoculated with koji mold. It was fermented between 3-7 days in dark vats and then fermented again with lactic acid and yeast for several months. The longer the fermentation period, the deeper the flavour became.

During the Han it was also adopted by Buddhist monks who wanted to avoid fish sauce. Soy sauce became popular with Chinese Buddhists to enhance their bland vegetarian diet. By the 8th century CE soy sauce had spread to Japan and Korea where it received further refinements.

Tea

According to legend, the tea culture began around 2737 BCE when the leaf of *Camellia sensis* fell into the boiling pot of water of Emperor Shennong. Tea originally was used for medicinal purposes as a stimulant, digestive aid and treatment for fever. Before the Qin Dynasty (221 BCE) and into the Han Dynasty, tea was consumed as a thick soup, boiled with millet until it became like porridge. Tea soup was a popular breakfast beverage used to provide energy for a hard day of work in the fields.

Powdered tea did not become widespread until the Tang Dynasty in the 8th century with the publication of Lu Yu's '*The Classic Art of Tea*." This book covered all aspects, such as how to cultivate tea plants, brewing techniques and the art of the tea ceremony. He promoted the idea that tea should be consumed mindfully without strong spices added. Infused tea leaves also originated during the Tang Dynasty.

Song Dynasty tea making painting

Cuisine

Imperial cuisine was well developed during the Zhou, with abundant cereals, vegetables and meat. The "*Book of Songs*" mentions the menu of dishes served at the imperial court banquet as well as the strict etiquette imposed.

- Fruits: apricot, pear, sweet crab apple, persimmon, melon, cherry, orange, tangerine and shaddock
- Sources of animal protein included ox, sheep, dog, pig, horse, deer, bear, wolf, elephant, chicken, pheasant, wild goose, turtle, snake, carp and shark.
- More than 30 kinds of common vegetables, rice, millet and beans
- Drinks were known as the 6 clears- water, vinegar and wine.
- The 5 quis which were wine residues made from rice and cereals.
- The 3 juis were filtered wines with the residue removed.

According to the "*Rites of the Zhou Dynasty*," an Emperor's banquet was comprised of 6 cereals, 6 animals, 6 clears for drink, 120 delicacies, 8 dainties and 120 urns of sauce. Senior officials had 20 more delicacies than junior officials, whose non-staple foods included roast beef, beef cooked in soy sauce, minced beef, roast sheep, roast pig, pork cooked in soy and minced fish.

REFERENCE: "The History of Chinese Imperial Food," http:/china.org.cn/English/imperial/25995htm#

The relationship between food and politics was very important during the Zhou, Qin and Han Dynasties (1122 BCE-220 CE.) Good chefs were highly prized and often entered politics, such as Yi Yin, Peng Zu, known as the founder of Chinese cuisine, and Yi Ya.

Zhou imperial households employed more than 2,300 staff in 22 departments responsible for growing, transporting, preparing and tasting food, service, water management, wine officers, nutritionists and assistants. Banquets and feasts were expected whenever the Emperor met with princes or dukes.

During the Spring and Autumn period the great philosopher **Confucius** wrote about many aspects of correct living, including the consumption of food. He believed that freshness is most beneficial to the human body. Stale, overcooked or rotten food should never be consumed.

- Simple fresh food with good quality rice should be the main staple.
- All meat should be cut into small pieces and eaten with chopsticks.
- A balanced diet should consist of rice, meat and vegetables.
- Food should not be overindulged during festive seasons.
- Ginger prior to eating improves digestion.
- Wine must not be brought from unwholesome places.

Food culture flourished during the **Han Dynasty** (206 BCE-220 CE) with various books appearing, including the "*Book of Foods*" by Cui Hai, "*The Seven Advices*" by Mei Cheng and parts of "*Essentials for the Common People*" by Jia Sixie. The technique of fermentation to make staple foods such as steamed cakes, stuffed buns, steamed buns and soy sauce was already fully adopted during the Han Dynasty. Han imperial kitchens grew vegetables in hothouses, so they were not limited by the season.

Noodles became a stable during the Eastern Han Dynasty (206-220 BCE) and were made of wheat. However, the oldest noodles discovered at Lajia were made of millet 4,000 years ago. Wheat gradually replaced millet as the favoured cereal crop throughout China by the Han. Noodle shops remained open all night during the Song Dynasty (960-1279.)

After the Han, thick soup became less important and roasted meats were eaten only when people drank wine. Stir-frying dishes in woks became the chief cooking method for all classes during the **Southern and Northern Dynasties** (420-589 CE) Buddhism was spreading in China and vegetarian dishes gained popularity, along with gluten.

During the **Sui** and Tang Dynasties (581-907 CE) there were more varieties of food and greater attention was paid to flavour, taste, colour and presentation. Famous imperial dishes still eaten today included fried prawns, crab rolls, phoenix cakes, crystal dragon and steamed Mandarin fish without soy sauce.

The **Tang Dynasty** brought together many regional cuisines from its large empire, including northern Vietnam and Inner Mongolia. Cooking methods included boiling, stir-frying, deep frying, roasting steaming and stewing. Due to the expansive trade networks along the Silk Road, many items from western Asia and Europe were added to the cuisine. Ginger was used extensively in recipes. Many foods were fermented, pickled, salted or brined for preservation.

- Vegetables- root crops, leafy greens, soybeans and other legumes.
- Fruits- bananas, peaches, pomegranates, pears, oranges and crab apples, jujubes, apples, apricots, pears.
- Meats eaten by all classes- pork, mutton, lamb, waterfowl and turtle. In the south fish, crabs, shrimps, oysters, jellyfish, puffer fish and clams were favoured.
- The elites ate delicacies like sea otters, monkeys, elephants, marmots, birds' nests, ants' eggs and unborn mice stuffed with honey.
- Beverages- ales, flavoured water, wine, fermented milk and of course, tea which often came in small bricks. Green and black were the most common teas.
- Imported foodstuff- dates, pistachios, figs, pine nuts, mangoes and sugarcane which was brought to China from India during the Tang.

Recipe from the Sui Dynasty: Egg fried rice (581-618 CE)

- Ingredients: 600 g leftover cooked white rice

- 4 large eggs
- 4 stalks green onions, or shallots
- 2 tsp salt, ½ tsp white pepper. 5 tbsp oil
- Method: 1. Loosen cold cook rice.
- 2. Beat 4 eggs and pour onto the rice, mixing the rice until it has been coated with egg.
- 3. Heat wok to 175 degrees C or 350 degrees F and coat the wok with oil.
- 4. Spread out egg in wok and fry for 10 seconds.
- 5 Stir-fry until rice is fluffy and egg is cooked.
- 6. Add shallots, salt, pepper and stir-fry until evenly distributed.
- 7. Add 1 tbsp of oil and mix. Optionally add other vegetables and meat.
  https://auntieemily.com/ancient-chinese-egg-fried-rice/

Ancient Chinese banquet

Sericulture, or silk making, was exclusively produced in China for thousands of years. Cocoons of silkworms have been uncovered at sites dating from 5,000 years ago. Shang Dynasty inscriptions on bones contained the characters for silkworm, mulberry and silk.

According to legend, the Lei Zu, wife of the Yellow Emperor Huangdi, was sitting under a mulberry tree enjoying her tea when a silkworm cocoon fcll into her bowl. She realised that the unwinding thread could be used to make yarn, and for that discovery she became known as the "Silk Goddess."

By the 5th century BCE, silkworms were bred in six provinces, with Hangchow the centre for sericulture. Initially worn only by the members of the imperial family, silk was worn by all classes and used as payment for taxes by the Han Dynasty. During the Han, the quality of embroidery and dyeing greatly improved. Silk was also carried along the Silk Road to the Roman Empire, but its production was kept secret. Any attempt to smuggle the eggs, caterpillars or moths was punished with the death penalty.

Mulberry leaves have been used in Chinese herbal medicine and the berry was used as a tonic by emperors and named the "sacred folk fruit." In the "*Book of Songs*" the uses for mulberry leaf, bark (papermaking) and fruit were mentioned.

Image-silk production

The Silk Road formally opened up China to the western cultures and cuisines in 125 BCE during the Han Dynasty. Prior to this important event, very limited trade between China and Eurasia had occurred, but overall, the vast Gobi desert had been an effective barrier to trade. During the Han:

- Crops like carrots, walnuts, garlic, flax and cucumber, eggplant, pomegranates and onions were introduced from Eurasia via the Northern Silk Road.
- Pepper and Indian spices were also introduced at this time. Originally used for medicinal purposes, pepper was cultivated in China during the Tang Dynasty and used in cooking.
- Grape seeds were first imported into China. Many Eurasian cultures had been making wine since Neolithic times, but for the Chinese the idea was novel.
- Many foods unknown to the western world passed along the Silk Roads from China to the Roman Empire, including

apricots, peaches, citrons, pistachios and rice. Chinese inventions like paper and porcelain also made it to Europe.

During the Tang Dynasty flat bread and cakes were brought to Chang'an from Central Asia. *Biluo*, a translucent savoury cake became very popular as western Asians settled in China.

China initially traded silk for horses from central Asia which were large and swift, making them prized warhorses. The Mongols ruled China from the 13th to 14th centuries and introduced many dairy and fermented dairy products from mare's milk.

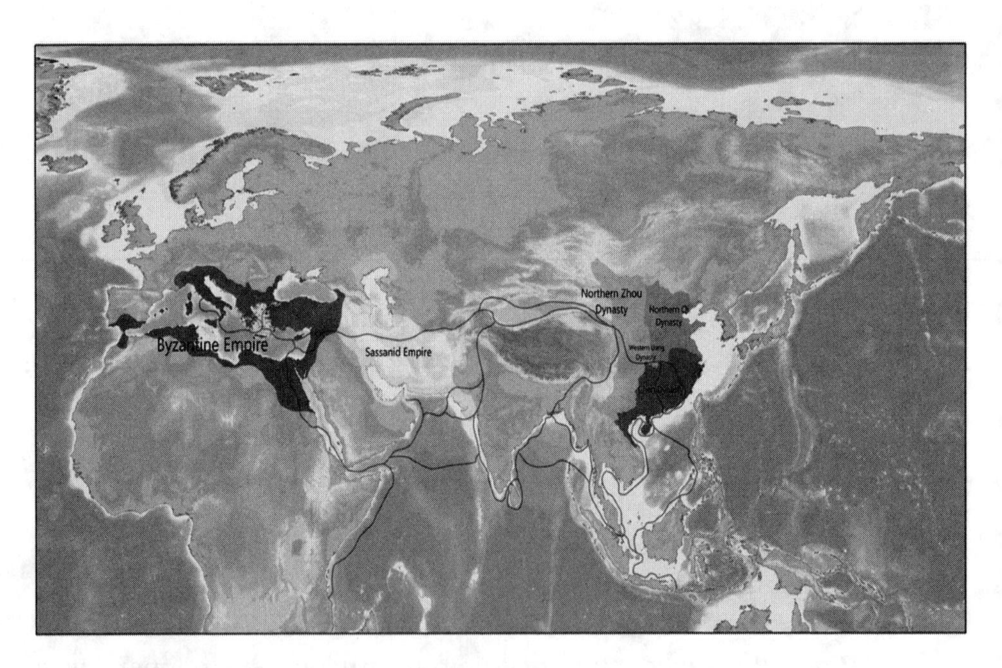

Silk Road routes, credit Aldan-2 CC BY-SA 4.0

Medicinal herbs have been used in China since the Neolithic age. The "*Huangdi Neijing*" was believed to have been written by the semi-mythical Emperor around 2600 BCE, but probably dates from around 300 BCE. The "*Shennong Jing*" (Classic of Shennong) is the oldest book on pharmacology and dates to the Han Dynasty. It has descriptions of about 365 curative herbs. Books on acupuncture and moxibustion appeared over the following centuries. In 657 CE, Emperor Gaozong of the Tang Dynasty commissioned an official *materia medica* with properties of 833 plants, herbs, vegetables, fruits, cereals, stones and minerals.

Aquaculture began around 5,500 years ago in China with the farming of the indigenous common carp. Carp were grown in ponds on silk farms and fed silkworm nymphs and faeces. The earliest known treatise on fish farming written by Fan Li in 475 BCE was called "*Yang Yu Ching*" (Treatise on fish breeding.) It was banned during the Tang Dynasty because the name for common carp sounded like the Emperor's family name. However, this ban led to polyculture, with five different species of carp cultivated. From 1500 CE, carp fry were collected from rivers and reared in ponds.

CHINA takes food security very seriously. In 2013, it adopted a national strategy for food security based upon self-sufficiency of domestic grain production, moderate imports and technological support. Cutting edge technologies such as biotechnology, gene editing, big data and AI are widely used in agriculture.

China is increasing the number of reservoirs to ensure an increasingly large percentage of arable land is well irrigated. To make the fishing industry sustainable, the Chinese have introduced strict conservation laws, by expanding the scale of fish farming and placing moratoriums on fishing in the Yangtze River,

> "Thanks to its food security strategy, China has witnessed bumper harvests year after year, with abundant production of vegetables, fruits and milk products. Moreover, food prices have remained generally stable in the country. And when food prices increased in the global markets earlier this year, China decided to cut grain imports, reducing the chances of market volatility and contributing to global food security."

REFERENCE: Li Guoxiang, “Country takes great strides in food security,” “China Daily,” October 15, 2022

https://global.chinadaily.com.cn/a/202210/15/WS6349fa73a310fd2b29e7c944.html

Modern China

KOREA

Agriculture spread to northern Korea from northern China with the introduction of foxtail and broomcorn millet after 8000 BP. Millet production appeared in southern Korea around 5,500 years ago during the Middle Chulum Neolithic period. Soybean, millet and adzuki were available at this time, whereas rice, barley and wheat arrived in the peninsula later.

By the Bronze Age in 1500 BCE, rice was becoming an important crop. Plant remains from Sangdong-dong and Songguk-ri show that rice was eaten by all segments of society. Other crops grown were red beans, soybeans and millet. During the Mumun Period from 1500-300 BCE, agricultural societies began, with extensive dry-field and paddy-field rice production. However, during the first three centuries CE, it became a luxury food used for tax payments.

The Three Kingdoms (4th-7th centuries CE) AND Unified Silla (C7th-10th CE) periods in Korea led to agricultural innovations such as expansion of iron tools, implementation of field systems and establishment of large-scale irrigation schemes for wet-rice farming.

Cuisine

During the Mumum period animals were domesticated, and the fermentation of beans was developed. The Baekje kingdom (18 BCE-660 CE) was known for its cold foods and fermented foods like kimchi. The spread of Buddhism in the 4th century CE brought in more vegetarian dishes. Rice became popular in the Silla and Baekje Kingdoms of southern Korea. The most traditional method of cooking the rice was to use an iron pot called a sot beginning in the Silla period.

Excavations at Okbang in South Gyeongsang province indicate that soybeans were cultivated as a food crop 3,000 years ago. Soybean sprouts were cooked as a vegetable, while dubu (tofu) was cooked for its protein.

Pork, chicken, beef, dog, fish and seafood have all been consumed since the Bronze Age. Pickled and fermented vegetables were mentioned in the historical record of the Three Kingdoms, and **kimchi** became popular during the Silla period. It was made during the winter by fermenting the vegetables and burying them in the ground in brown ceramic pots called onggi. Radishes were the most popular vegetables, along with the addition of various spices and seasonings, although chilli spice did not exist until the 17th century when it was introduced to Asia by Portuguese traders. Cabbage was also used for kimchi during the Three Kingdoms.

A poem on the radish by Yi Gyubo from the 13th century:

Pickled radish slices make a good summer side-dish,
Radish preserved in salt is a winter side-dish from start to end.
The roots in the earth grow plumper every day,
Harvesting after the frost, a slice cut by a knife tastes like a pear.

—Yi Gyubo, Donggukisanggukjip (translated by Michael J. Pettid, in Korean cuisine: An Illustrated History)

Some Chinese food historians claim that kimchi is a version of Chinese paochai, but Koreans reject this claim and see it as an attack upon their national dish.

Pottery in Korea dates back to the Neolithic period, when brown earthenware bowls with incised decoration showed a cultural link with communities in the Liaoning province of China 10,000 years ago. Jeulmum pottery, "comb pattern pottery" was created after 7000 BCE in the central regions of the Korean peninsula.

Mumun pottery was discovered at dolmens dating from 3000 to 400 BCE. They were used for practical and ritualistic purposes. Jars with handles on both sides were developed as rice cultivation spread. During the Samhan period, pottery was fired at around 900 degrees C and named Gimhae, after the discovered site.

During the Three Kingdom and Silla periods, pottery was technically very advanced, with celadon glazed wares introduced from China. This glaze, also known as green ware, is produced by firing a glaze containing a little iron oxide at a high temperature in a reducing kiln.

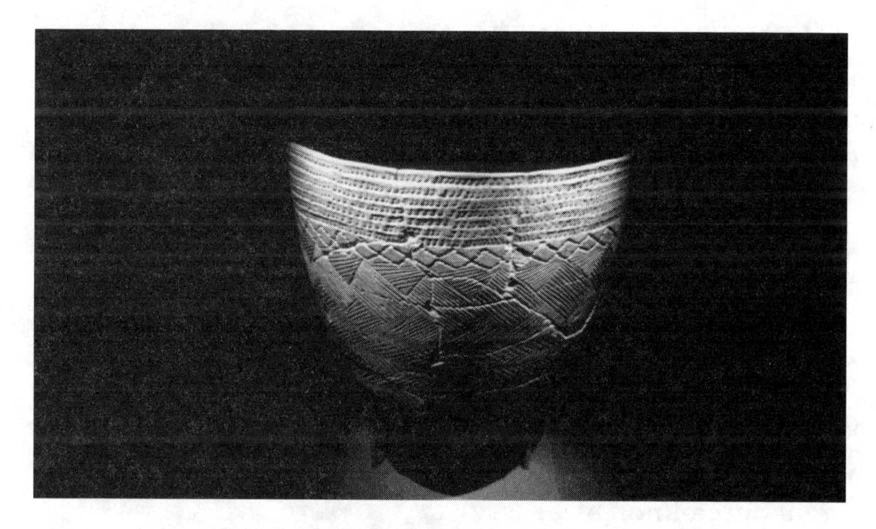

Neolithic pot

KOREA is a peninsula divided into prosperous South Korea and communist North Korea, both with entirely different economies. North Korea has a current grain shortage of 1.35 million tons, but because of economic sanctions has few trading partners. The regime's priority in making NK a defence economy has impacted food production and living standards. The shortage of grain in 2021 is believed to have been brought about by extensive flooding, the pandemic and international sanctions.

A typhoon in 2020 impacted grain production by almost 860,000 tons, equivalent to the loss of two months of food grain. Insufficient agricultural production, antiquated machinery, continuous mono-cropping and climate disasters have resulted in chronic food insecurity and malnutrition for up to 40% of the population.

REFERENCE: https://www.orfonline.org/expert-speak/explaining-north-koreas-food-crises-in-the-context-of-food-security-and-sustainable-food-systems/

JAPAN is an archipelago with semi tropical conditions in the south and very cold winters in the northern island of Hokkaido. Being close to Korea and China, its food was heavily influenced by those cultures. The Neolithic **Jomon** Period, from 14,500-300 BCE saw the introduction of pottery and cereal grasses such as millet. The first crop cultivation appeared around 5700 BCE with slash-and-burn agriculture, and intensified around 4000 BCE. Fleshy fruits and nuts were also cultivated in Jomon communities. Rice was introduced from the mainland around 1250 BCE, but its cultivation was not widespread until around 800 BCE. Wet-field rice cultivation appeared two centuries later when immigrants brought the technique to the south-west, from where it spread northwards.

The **Yayoi** Period (300 BCE-250 CE) saw Chinese immigrants bringing soybeans, wheat, barley, hops, bottle gourds, peaches, persimmons and adzuki beans to Japan. During this period of intensive agriculture, bronze and iron tools from Korea in the **Kofun** Period (250-538 CE) were introduced. Wooden hoes and spades were replaced with metal, and irrigation spread. In the northern island of Hokkaido dry field rice cultivation was practised due to decreased rainfall.

Over the next few centuries, agriculture did not improve for the average farmer and irrigation techniques were insufficient to prevent crop failures and famine. Under these poor conditions, most farmers preferred to work for landed aristocrats on their estates which were properly irrigated. Independent farmers had to settle for dry fields like millet, barley, hemp and wheat, while rice was grown and reserved to pay taxes.

Crops, especially rice, had their own deities. Inari was the national rice god, along with the protective Shinto spirit *Ta no kami.* Religious ceremonies and rituals were enacted around sowing and harvesting time to ensure a good crop protected from disasters. Rice ceremonies were particularly important and involved the Emperor.

CUISINE

Seafood has always been more popular than meat in Japan, including shellfish, seaweed, sea cucumber, eel, carp, sea bass, mackerel, sardine, salmon, prawns, squid, jellyfish and crab. Fish were also dried and transported inland. The introduction of Buddhism into Japan during the 6th century CE, with its avoidance of killing animals and birds, was another reason vegetables and seafood dominated the diet. In 625 CE, Emperor Temmu banned the consumption of cattle, wild animals such as cow, dog and monkey included, although fallow deer and wild boar were allowed to be eaten.

Rice was boiled, steamed or cooked and dried. It was mixed with vegetables to make rice cakes or a thick porridge. Rice wine, sake, appeared during the Yayoi era but did not become an industry until the 12th century. Vegetables included red beans, bamboo shoots, cucumbers, onions, spring onions, Japanese sweet potatoes, yams and radishes. Food was seasoned with salt, ginger, fish, garlic, vinegar, soy sauce and fish broth. The soybean was a versatile legume which could be made into miso, a flavouring paste, tofu, (bean curd) or soy sauce.

Ancient Japanese grew many fruits such as raspberries, tangerines, persimmons, loquats, plums, apples and strawberries. Mulberry trees were grown to provide nutrition for silkworms. Chestnuts, walnuts and pine nuts provided oils, as dairy was uncommon except for the elite classes.

Tea introduced from China in the 7th-8th CE by the monk Erichu was associated with the sage Daruma, the founder of Zen Buddhism. It was adopted by Zen Buddhist monks who used it as an aid to ward off sleep during meditation. The tea was prepared by pounding the leaves and making a ball with ginger or amazura, pressed from wild grapes, and left to brew in boiling water. From 1200, during the **Heian** Period, tea rooms and tea schools were opened, with luxurious tea-tasting parties gaining popularity amongst the samurai and elite.

During the 12th century, a monk named Elsai introduced a specific way of preparing tea using powdered green tea. This matcha was placed into a bowl which was filled with hot water. Tea rituals became common among Buddhist monks. Centuries later, the traditional tea ceremony was introduced by Sen no Rikyu whose book "Southern Record" explained his knowledge of tea and its preparation.

The **Nara Period** from 710-794 CE saw the mastery of fermentation with the introduction of natto and unleavened bread as well as *miso, mirin* and *shi.* The Heian Period (794-1185 CE) is when literature flourished, with references to aristocratic dinner parties held in the pleasure palaces of Heiankyo (Kyoto.) Aristocrats ate two meals a day, one at 10 am and the second at 4pm. Kaiseki, ceremonial feasting, began during the Heian era with up to 28 dishes in four categories: dried foods, fresh foods, fermented foods and sweets. Fish and chicken could be either raw or cooked. Salt fermented fish or jellyfish were also served.

The earliest form of **sushi,** called *narezushi,* was created during the Yayoi period, with salted fish coated in fermented rice which prevented it from spoilage. The rice was discarded when the fish was eaten. After the 8th century CE, *haya-zushi* was assembled so the rice and fish could be eaten at the same time. The rice was no longer fermented but mixed with vinegar, vegetables and dried ingredients

Noodles were brought to Japan from Song Dynasty China at the end of the Heian period to the beginning of the Kamakura period from 1185-1333. Monks cultivated the wheat and grinding objects. Udon noodles were first mentioned in a document dated to 1347, and soba noodles, made from buckwheat appeared in 1438.

RECIPE: *So* dessert was a dairy product consumed by the nobility from the Asuka Era (538-710 CE) to the Heian Era which was referenced in the "Engishiki" book of 927.

Method: "Boil down 1 to (an ancient unit of measurement; around 18 liters/4.76 gallons) of milk to get 1 sho (an ancient unit of measurement; around 1.8 liters/0.48 gallons) of so."

During the pandemic lockdown of 2020-1, there was an excess of milk in Japan, so Japanese looked to this ancient recipe in order to use large quantities of the dairy product. The only ingredient of so is milk which has to be heated and stirred until it turns into the consistency of cottage cheese, which takes about one hour. Once it reaches this stage, it can be kneaded until it resembles hard cheese and chilled for two hours. Overall, the result is a sweet dessert which can be cut into slices and embellished with honey or soy. While *so* is a one ingredient food, it is very labour intensive as it has to be continually stirred for over an hour. This intensive procedure and the rarity of dairy in medieval Japan explains why it was only consumed by the elite classes.

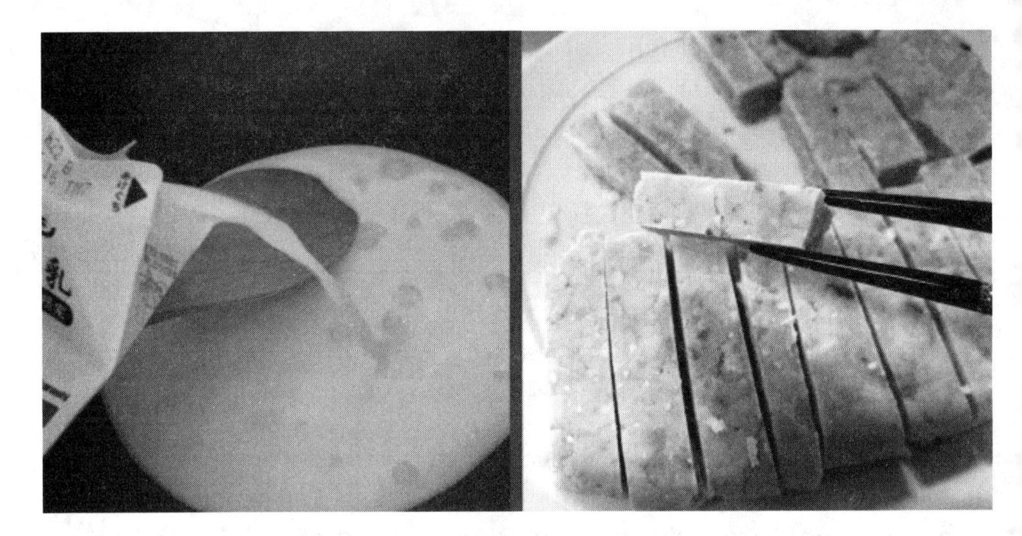

Reference and image: Casey Baseel, “How to make so, Japan’s 1,000 year old dessert recipe that’s back in fashion,” “Sora News 24,” March 13, 2020. https://tinyurl.com/mry39aju

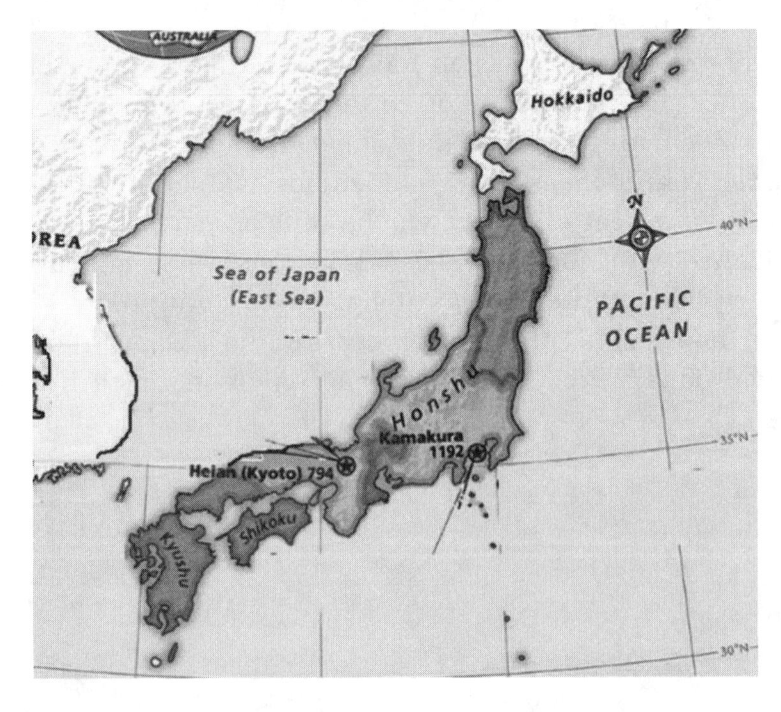

Ancient and medieval Japan

Jomon Pottery is considered to be the oldest Neolithic pottery in the world at 16,000 years old. Early production sites have been found around Shinonouchi in Nagano (13,000 BCE) and Odai-Yamamoto in Aomori (14,500 BCE.) The oldest corded ware vessels were most likely fired at 600-900 degrees C in open pits, as kilns have not been discovered.

Most Jomon pots had rounded bottoms and were used to boil food or water. The corded impressions were created by pressing rope into the clay before it was heated.

The Jomon era lasted until 300 BCE when many migrants from Asia arrived, bringing in new pottery techniques, forms and decoration. The Yayoi Period lasted until 250 CE and was marked by finer pottery of a reddish colour with no decoration. Sue stoneware was introduced from Korea in the Kofun period and was of a higher quality. In the 3rd-4th centuries CE, the anagama kiln, a tunnel kiln on a hillside, and the potter's wheel appeared in Kyushu from the Korean peninsula. The Sue pottery was fired at temperatures over 1200-1300 degrees C, and became elite tableware during the Heian period.

10,000 year old Jomon pot

Most ceramics during the Heian period had simple green lead glaze, three colour lead glaze was introduced in Japan from Tang China in the 8th century.

Atsuma pot, Heian burial ground

PAPUA NEW GUINEA is one of the only places where early agriculture was undertaken which did not lead to more sophisticated societies. Around 12,000 years ago, the island was attached to mainland Australia, forming part of the continent of Sahul. The earliest form of farming at Kuk was swidden (slash and burn) but eventually the farmers experimented with more intensive forms of cultivation, including raised beds and drainage canals.

Kuk Early Agricultural Site, listed with UNESCO, consists of 116 ha of swamps at 1,500 metres above sea-level in the western highlands. Although the landscape has been almost continuously farmed or up to 10,000 years, modern farming activities do not intrude upon the archeological diggings.

In 1966, archeologists discovered evidence of early agriculture and irrigation at Kuk Swamp in the Highlands dating back to 10,200-9910 BP. Evidence for planting and tending crops such as banana, taro and yam dates back to 6590-6440 BP, and irrigation around 4350-3980 BP. The man-made irrigation features include channels associated with natural levees near an ancient waterway, with evidence of digging and tethering plants in a cultivated plot nearby.

During Phase 2, (6950-6440 BP) the inhabitants created circular mounds for planting crops in what is known as raised field agriculture. By Phase 3, (5350-2800 BP) the farmers had constructed a network of drainage channels to drain water from the productive soil to the swamplands to facilitate farming.

Stone tools such as flaked scrapers, mortars and pestles were examined by researchers who found the phytoliths of taro, yams, grasses, and palms. The inhabitants also supplemented their diet with protein from hunting animals.

REFERENCE: “Kuk Swamp: Early Agriculture in Papua New Guinea,” Thought.Co https://www.thoughtco.com/kuk-swamp-early-evidence-for-agriculture-171472

According to UNESCO: “It is an excellent example of transformation of agricultural practices over time from mounds on wetland margins around 7,000-6,400 years ago to drainage of the wetlands through digging of ditches with wooden tools from 4,000 BP to the present.”

UNESCO: Kuk Early Agricultural Site
https://whc.unesco.org/en/list/887

The Lapita were Austronesian people who left ceramic objects on Pacific Islands like New Caledonia, Fiji, Tonga and Samoa, the Solomons and Vanuatu over 3,600 years ago. Lapita pottery was produced between 1600 and 1200 BCE on the Bismarck Archipelago, and 1000 BCE in Fiji. Typically, the pottery was low-fired earthenware, tempered with shells or sand and included cooking pots, bowls, and beakers.

The Lapita domesticated dogs, pigs and chickens. They grew tree crops like coconuts, bananas and breadfruit as well as root crops like yam and taro. Other protein sources were derived from the oceans, including fish and molluscs. Many of the material culture elements appear to be South East Asian in origin, such as paddy field agriculture, stilt houses, outrigger boats and especially pottery. According to DNA evidence they had descended from inhabitants of Taiwan and the Philippines who sailed to these islands. No evidence has been found on many islands of settlements in earlier agricultural stages.

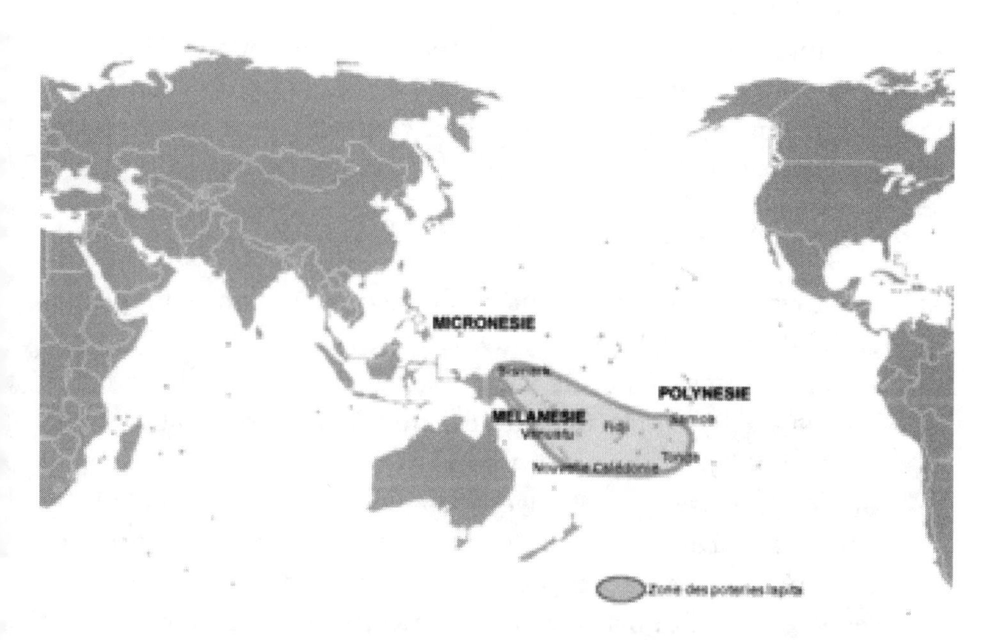

Regions where Lapita pottery has been found. Credit Christophe cage CC BY-SA 3.0

Lapita pot

HAWAII may have been the first place to practice mariculture, seawater farming, around 1,500 years ago. The ancient Hawaiian fishponds were part of an integrated subsistence/bartering economy that included agriculture, aquaculture and animal rearing.

Ancient Hawaii had a highly stratified society with chiefdoms and even kings who owned all land including fishponds and fields. Much of the land and fishponds were contracted to lesser chiefs who leased them out to families. When Captain Cook reached the Hawaiian Islands in 1770, at least 360 fishponds existed.

The Hawaiian fishponds were divided into four categories:

- Freshwater taro fishponds stocked with mullet, perch, milkfish and freshwater prawns
- Other freshwater ponds were usually inland ponds or lakes excavated by hand for mullet, milkfish, perch. They were harvested by woven reed nets.
- Brackish water ponds were coastal ponds excavated by hand, or formed by piling mud, sand or coral to form an embankment with a canal constructed to catch tidal seawater. With spring water often mixed in, they were very productive.
- Seawater ponds had a long wall constructed of coral or lava rock. At least 22 species of marine life flourished in these ponds which were common on Molokai and Oahu. Canals were constructed into the walls of the ponds for stocking, harvesting and cleaning the ponds, with wooden grates which allowed small fish to enter. Large fish were harvested by nets near these grates as they tried to escape.

Some of these fishponds were large in size. The largest pond at Mauna Lania, Lahuipua'a is 5 acres in size and reaches a depth of 18 feet. It may have been built as far back as 250 BCE. A brackish pond, known as the Menehune Alekoko firshpond near Lihu'e Kaua'I was built by the legendary pygmy race known as the menehune, according to local legend.

Alekoko fishpond, credit Collin Grady. CC BY-SA 2.0

Poi is a staple Polynesian/Hawaiian food made from breadfruit, taro, plantain or starchy vegetables which can be eaten fresh, or sweet and sour when fermented. It was traditionally made by mashing cooked starch on a wooden pounding board with a carved pestle made from wood, coral or basalt and adding water. It is believed to have originated in the Marquesas Islands before spreading to eastern Polynesia, including the Hawaiian Islands. Hawaiians still cook taro for hours in an underground oven called an imu. Natural fermentation occurs days later due to the lactobacillus bacteria, yeast and *Geotrichcum* fungus.

Poi pounders at the Kauai museum. Credit Richard Horne CC BY-SA 4.0

AUSTRALIA, until recently, was believed to be devoid of any ancient agricultural practices, but research by alternative writers such as Bruce Pascoe and Bill Gammage has challenged this conventional view. Australia is the driest continent on Earth, with climatic zones ranging from alpine to desert and rainforest. The aboriginal tribes, with their knowledge of botany and their ability to locate water in any environment, were able to thrive for over 60,000 years by primarily hunting and foraging.

Bill Gammage, who wrote "The Biggest Estate on Earth: How Aborigines Made Australia," uncovered a complex system of land management using fire, the life cycles of native plants and the natural flow of water to ensure plentiful plant foods and wildlife throughout the year. Furthermore, this management ensured that these food sources would be protected for other wandering tribes. The aboriginals firmly believed that the beings from the Dreamtime had given them all the knowledge and laws needed to provide them with food, and they were tasked with managing and protecting the land.

In his book "Dark Emu," Bruce Pascoe argues that across the continent, Aboriginal tribes were domesticating plants, as well as sowing, harvesting, irrigating and storing them. Many of his examples come from the sources of early explorers and settlers, such as Thomas Mitchell who reported that native grasses in the Darling River area were pulled and piled in large stacks while they were still green and full of seed. When the seed ripened, the heap was threshed and the seeds were gathered. Grinding stones used for making flour out of millet have been found across New South Wales. Some have been dated to 36,000 years, indicating that flatbread made from yam, daisy and kangaroo grass is the oldest in the world.

The diet of Aboriginal people was varied, especially those living near rivers. They ate fish, shellfish, small mammals, reptiles, roots, yams, tubers, birds and emu eggs for thousands of years. Native plants such as macadamia, quandong, lemon myrtle, Tasmanian pepperberry, Bunya nut and finger lime were also harvested, as were native berries such as minya. Yams provided necessary carbohydrates and were widely harvested by leaving the top of the tuber attached to the plant so that the tubers regrew. Bunya nuts come from giant pine cones which are filled with dozens of nuts and are an important food source for Queensland tribes. Today these foods are affectionately referred to as "bush tucker."

Polish-German zoologist Johann Blandowski, who explored the Murray Darling region of NSW from 1856-7, sketched many images of the fauna, flora and people of the region. One image, now in the British Museum, shows clay ovens which were used to cook bulrush roots collected by the women in shallow swamps between January and March. The larger roots were tied into bundles and roasted in clay ovens for 2-3 hours.

Unfortunately, Blandowski had a falling out with the Philosophical Institute in Melbourne and eventually returned to Europe, where his sister published his sketches after his death.

Photo from William Blandowski of clay ovens, now in British Museum

Although aspects of "Dark Emu" have been criticised by academics, it cannot be argued that Aboriginal people in southern Australia have created hydraulic complexes for eel harvesting and fish traps. In south western Victoria, large wetlands were the feeding areas for native eels which returned to the sea to breed in autumn. The Aboriginal people constructed an elaborate system of traps and canals at places like Lake Condah where stone canals and walls up to one metre high and more than 50 m long were built from volcanic rocks. Apertures were built into walls to place eel pots, nets and traps which were built on different levels. These eel trap systems covered up to 100km2 and contained weirs, channels and dams as well as areas for smoking and were at least 8,000 years old.

One colonial observer near Mt William wrote:

> At the confluence of this creek with the marsh observed an immense piece of ground trenched and banked...which on inspection I found to be the work of the Aboriginal natives,

purposefully constructed for catching eels... These trenches are hundreds of yards in length. I measured in one place in one continuous trepple line for the distance of 500 yards. These treble watercourses led to other ramified and extensive trenches of a most tortuous form. An area of at least 15 acres was thus traced over...These works must have been executed at great cost of labour...the only means of artificial power being the lever...This lever is a stick chisel, sharpened at one end, by which force they threw up clods of soil and thus formed the trenches, smoothing the water channel with their hands. The soil displaced went to form the embankment...

"The plan or design of these ramifications was extremely perplexing and I found it difficult to commit [to] paper, in the way I could have wished, all its various form and curious curvilinear windings and angles of every size and shape and parallels, etc. At intervals small apertures left and where they placed their arabeen or eel pots. These gaps were supported by pieces of the bark of trees and sticks. In single measurement there must have been some thousands of yards of this trenching and banking. The whole of the water from this mountain rivulet is made to pass through this trenching where it reaches the marsh; it is hardly possible for a single fish to escape. I observed a short distance higher up, minor trenching was done through which part of the water ran it course to the more extensive works. Some of these banks were two feet in height, the most of them a foot and the hollow a foot deep by 10 or 11 inches wide. The main branches were wider.

"Around these entrenchments were a number of large ovens or mounds for baking, there were at least a dozen in the immediate neighbourhood..."

(Robinson, 9/7/1841 in Clark 2000b in the Report to Aboriginal Affairs Victoria).

These structures were forgotten and neglected until the 1970s, when Dr. Peter Coutts of the Victoria Archeological Survey of Lake Condah found extensive Aboriginal fish-trapping systems. They were comprised of hundreds of metres of excavated channels and dozens of basalt block dam walls, made from hundreds of tonnes of basalt. Furthermore, numerous C-shaped basalt block structures, measuring 3-4 metres across are believed to be house formations. Coutts believed the traps to be up to 3,500 years old, but radiocarbon dating of tiny charcoal fragments with sediments yielded a date of at least 6,600 years old.

Harry Lourandos, PhD, investigated a huge fish trap at Toolondo, 110 km north of Lake Condah. Up to 2.5 m wide, and over a metre deep, this earthen channel, 3 km long, moved eels into a swamp.

In the early 2000s, Heather Builth, PhD, worked closely with the Gunditjmara to create 3D computer maps of the fish traps, channels and basalt block walls along the southern end of the **Budj Bim** landscape. She discovered these stone features were constructed across the lava flow from Mt Eccles to form a complex system of artificial ponds to hold floodwaters and eels at different stages of growth. She described the network of ponds as aquaculture.

In 2018, Prime Minister Turnbull announced that the federal government had included the Budj Bim cultural landscape on its World Heritage Tentative List, and subsequently, it was added to the UNESCO World Heritage List.

In early 2020, devastating bush fires ravaged parts of the area, revealing extra sections of the complex which had been covered by vegetation. The fire uncovered another smaller system including a 25-metre channel which was only 20 m off the walking track.

In Brewarrina, NSW, ancient stone fish traps, known as **Baiame's Ngunnhu**, were built thousands of years ago by the Wailwan people. The stone traps consist of a series of dry-stone weirs and ponds arranged in the shape of a stone net across the Barwon River.

The structures occupy the entire length of a 400m-long rock bar that extends from bank to bank across the river bed. These rocks are locked in together, with large stones placed along the tops of the walls.
W.C Mayne, Commissioner of Crown Lands at Wellington, described them:

> "In a broad but shallow part of the head of the River there are numerous rocks, the aborigines have formed several enclosures or Pens...into which the fish are carried, or as if were decoyed by the current, are there retained. To form these must have been a work of no trifling labour, and no slight degree of ingenuity and skill must have been exercised in their construction, as I was informed by men who have passed several years in the vicinity, that not even the heaviest floods displace the stones forming these enclosures."

Unfortunately the fish traps have been damaged since colonisation, and the construction of a weir upstream further damaged them, but efforts have been made by governments to preserve the traps which may be thousands of years old. Soon after the construction of the Brewarrina Weir in 1971, the Council employed local indigenous people to restore parts of the traps which became listed on the Australian National Heritage List in 2006.

The Brewarrina Fishery.

Lindsay Thomson, 1893

Bush tucker, Alice Springs Desert park: credit Tourism NT, travelnt.com

Although there are references in early European texts about Australian aboriginal people fermenting food and beverages, very few studies have been made. In 2016, Professor Vladimir Jiranek, of Oenology at Waite Campus, University of Adelaide, was recruiting students for a study on traditional fermentations of cider gum sap, Banksia and quandong roots from South Australia.

"There is an important historical and anthropological aspect to the project," Professor Jiranek said. "Some of these processes are not being practised anymore, but there may at least still be people with living memory of them. We're keen to record what we can while still possible."

"The work may also reveal that novel organisms are involved that are unique to Australia, making this an opportunity to identify some new species of yeast and bacteria, perhaps with interesting new properties. We will also characterise the composition of the saps, nectars and various extracts as fermentation substrates while the Australian Bioactive Compounds Centre will determine if bioactive compounds are also present," he added.

"Revealing the science of Aboriginal fermentation," University of Adelaide, October 21, 2016,
https://www.adelaide.edu.au/news/news88682.html

In 2020, Professor Jiranek announced that some of the bacterial and fungal communities from the cider gums they had studied could not be matched to existing databases.

"It also allows us to identify new strains, or species, of yeast and bacteria from the fermentations that are unique to Australia. Further work will characterise single microorganisms that have been isolated and grown from the cider gum.

"We are particularly interested in their fermentative abilities, their potential flavour impacts, how they've adapted to the cider gum environment and the possible symbiotic relationship they have with the trees."

Mills, Roby, "Uncovering the science of Indigenous fermentation," The University of Adelaide Newsroom," September 11, 2020, tinyurl.com/47ap6f6y

Mabuyag Island, located in the Torres Strait between Australia and New Guinea, has yielded evidence of banana farming 2,000 years ago. Fossilised traces of fruit, stone tools, retaining walls and charcoal were uncovered at the Wagadagam site in 2020. This is not so surprising, considering that banana cultivation occurred in New Guinea began at least 7,000 years ago.

Fox, Alex, "Traces of 2,000-year old Banana Farm found in Australia," "Smithsonian Magazine," August 18, 2020.

https://www.smithsonianmag.com/smart-news/2000-year-old-traces-banana-farming-found-australia-180975583/

PART 5

THE AMERICAS

CENTRAL AMERICA (MESOAMERICA)

OLMECS

MAYANS

AZTECS

SOUTH AMERICA

ANDEAN HIGHLANDS

COASTAL REGIONS OF PERU

AMAZONIA

NORTH AMERICA

MESOAMERICAN FOODS & SPICES—avocado, tomato, sweet potato, squash, maize (corn) beans, cacao, vanilla, sunflower

SOUTH AMERICAN FOODS—potatoes, quinoa, guinea pigs, passionfruit, pineapple, peanuts, cashews, acai, guava

NORTH AMERICAN FOODS—maize, squash, beans, turkeys, pumpkins

MESOAMERICA

Agriculture in Mesoamerica dates from 8000-2000 BCE; a period known as the **Archaic** period of Mesoamerican chronology. Prior to 8000 BCE, the area was populated by hunters/foragers who led a nomadic existence. During the Archaic period, major crops such as maize, (corn) beans and squash, often referred to as the "Three Sisters" were cultivated.

Maize was first domesticated in southern Mexico around 10,000 years ago in the Tehuacán Valley and spread slowly into the lowlands by 5600 BCE and on to Colombia between 7000-6000 BCE. Scholars believe that maize was domesticated from teosintes, a flowering plant in the grass family. Squash, which is related to today's pumpkin, was found in a cave dating back to 8,000 BCE. The bottle gourd which provided storage space for conveying water or collecting seeds, was also cultivated in the Archaic period. Other crops cultivated by 4000 BCE include manioc, chilli peppers, avocadoes, and later, tomatoes.

THE OLMECS, (1500-450 BCE) the earliest known culture of Mesoamerica, are also known as "The Rubber People" because they utilised the rubber trees growing in their region. They practised slash-and-burn agriculture to clear the forests for new fields which were located outside the village to cultivate maize beans, squash, cassava and sweet potato. Avocado, sunflower and cacao plants were available in the nearby forest. Rivers made it possible for them to create irrigation systems and harvest twice a year.

For protein they ate fish, turtle, snake, shellfish and molluscs as well as small mammals like peccary, opossum, rabbit and racoon. Larger animals killed for consumption were deer and domesticated dog, but Mexico lacked herding animals such as sheep and goats.

THE MAYANS were a cultural group centred in the Yucatan region of Mexico from around 1000 BCE until their collapse in about 900 CE. They were an agrarian society which overcame poor soils and unreliable water sources by building stone reservoirs for storing water, and practising slash-and-burn cultivation to add nutrients to the thin layer of topsoil. Mayan towns in the mountains created terraces for growing maize and other crops by lining them with a wall at the border to eliminate erosion. Irrigation canals were used to water the crops. In the swampy lowlands, the Mayan farmers created raised garden beds next to artificial canals. These beds were created by placing fertile mud with seeds from the bottom of the swamps onto reed mats, a process which yielded several crops per year.

Farming tools were made of stone and wood, as there was no Bronze Age in the Americas. Axes had a long shaft of wood with a sharpened stone attached. To plant the seeds they used wooden sticks.

Maize (corn) is a cultigen; human intervention is required for it to propagate. The earliest teosinte plants grew only one inch (25 mill) corn ears, with only one per plant. The Olmecs and Mayans cultivated numerous varieties of maize which they cooked, ground and processed through **nixtamalization**. As corn is a grain, it needs to be soaked and cooked in an alkaline solution like limewater, washed and then hulled before turning into cornmeal. Nixtamalized maize is more easily ground, flavoursome, aromatic and has fewer mycotoxins than unprocessed grain. The earliest evidence of nixtamalization is found in southern Guatemala, with processing equipment dating from 1500-1200 BCE. The Mayans and Aztecs developed nixtamalization using slaked lime and lye to create alkaline solutions which provided niacin to their diet. The spread of maize cultivation was accompanied by the nixtamalization process throughout the Americas.

Teosinte top, maize-teosinte hybrid middle, maize bottom. John Doebley CC BY 3.0

CUISINE

The Mayan diet was mainly vegetarian and included fruits/vegetables like avocado, pumpkin, squash, papaya, tomato, manioc cassava, sweet potato as well as local fruits like sour orange and mamey. Maize, the most staple food, was nixtamalized and baked into tortillas, tamales and flatcakes. **Tamales,** made with ground corn called masa, chillies and meat wrapped in a corn husk, have been eaten in Mexico for thousands of years. Steamed and fermented inside the corn husks or banana leaves, they were very popular as portable edibles for travellers. Because they are cooked without the use of ceramics, they are believed to predate the tortilla.

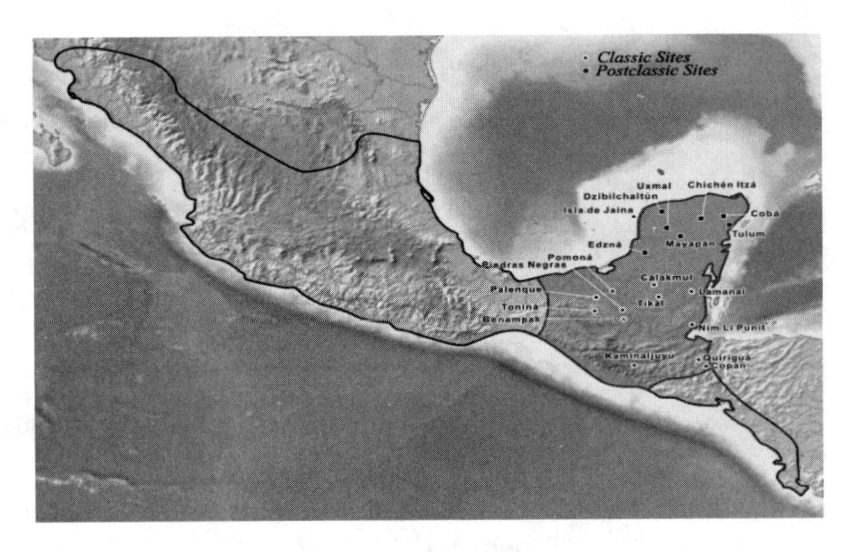

Mayan culture, credit Kmusser- CC BY-SA 3.0

The Mayans were among the first people to discover **cocoa** and make chocolate spiced with chili peppers and vanilla called *xocolatl*. They harvested the cocoa beans and fermented them for some time. Then they removed the shells of the beans, dried and eventually ground them into a paste or powder form. To make a chocolate drink, the Mayans poured it from a vessel at some height into another vessel several times so that the fats created a thick layer of foam on top. Only the rich and the elite were able to drink chocolate which could also be sweetened with honey. Another popular drink was alcoholic *balche* which was made by soaking the bark of the balche tree in honeyed water until it fermented. The result was a form of mead which was used in religious rituals. A favourite beverage was a corn gruel called *atole* which was sweetened with honey.

Mural with a chocolate beverage

Chili peppers (chilli, chile) originated in Bolivia and were first cultivated in Mexico 6,000 years ago. Cultivated independently in highland Peru, central Mexico and the Amazon, they were one of the first self-pollinating crops. They are part of the capsicum genus and *solanaceae* nightshade family.

Mayan cultivars

According to an article "Remembering the Future: How Ancient Maya Agronomists Changed the Modern World:"

"For more than 8,000 years Maya agronomists created cultivars or plant varieties of unequalled quality by combining science with selective plant breeding The goal was to develop cultivars that enhanced the lifestyle of their populace.

"After the discovery of America, Spanish explorers encountered Maya cultivars and they adopted these, disseminating them across the world. The adoption of the unique cultivars by peoples in Afro-Eurasia altered history.

"By the 1530 tomatoes were growing in Italy, maize was an African crop by 1590; papayas were grown in Asia by 1530, tobacco in 1520. In 1550 Europeans introduced cassava and peanut to tropical South East Asia and West Africa. This exchange of cultivars, animals and ideas became known as the Columbian Exchange...one of the events that established the modern world."

Mayan cacao goddess

World changing Mesoamerican/South American cultivars include:

Tobacco, cotton, turkeys, maize, sweet potatoes, tomatoes, peanuts

cassava, cacao, chicle, henequen, sunflower seeds, papaya, vanilla

chili peppers, beans, squash

Kon, J, "Remembering the Future: How Ancient Maya Agronomists Changed the Modern World," "Ancient Origins," August 28, 2017 https://www.ancient-origins.net/opinion-guest-authors/remembering-future-how-ancient-maya-agronomists-changed-modern-world-008695?

New World native plants. Clockwise, from top left: 1. Maize 2, Tomato 3, Potato 4, Vanilla 5, Pará rubber tree 6, Cacao 7, Tobacco.

THE AZTECS were a central Mexican culture which thrived from 1300 CE until the Spanish invasion in 1519. Maize was the most important staple, along with beans and squash, as well as New World varieties of the grains amaranth and chia. Proteins consumed were fish and wild game, including crayfish, iguanas as well as fowl such as turkeys and ducks. Various varieties of squash were eaten fresh or roasted including zucchini and capsicum. Tomatoes were often mixed with chilli in sauces or as fillings for tamales.

On the darker side, ritual cannibalism took place, as the gods demanded human sacrifice in order to sustain the world.

The main method of food preparation was boiling or steaming in two-handled clay pots called *xoctii* which were filled with food and heated over a fire. They could also be filled with some water to steam tamales on a small twig rack. Tortillas, tamales and casseroles filled with spicy sauces were the most common dishes, seasoned with salt and chilli.

Women did most of the cooking in a small kitchen with a triangular hearth. Food preparing implements were the *manos* and *mutates* for grinding nixtamalized grain, peppers and possibly cacao. The *metate* is a slightly concave stone slab for the grain which was ground by the stone mano. This tedious process of grinding the maize dough, *masa,* must have taken hours every day.

Apart from chili peppers, other herbs/spices included vanilla, allspice, Mexican oregano and Mexican coriander. Many alcoholic beverages were made from fermented maize, honey, cactus fruit, maguey (agave) and pineapple. *Octii* was fermented from maguey sap and consumed by all classes.

RECIPE- Tlahco (taco) made with traditional ingredients.

1. Corn tortilla
2. 1 pound of ground turkey
3. 2 large tomatoes, 1 onion, 3-4 serrano peppers, 2 poblamo chilis
4. 1 tbs dried epazote herb, pinch of salt, dried chilis

Method. 1. Core tomatoes and peppers and peel the onion.

2. Slice into large pieces and lay flatside down on a grill with low heat.

3. Turn with tongs occasionally until soft and charred.

4. Transfer charred vegetables to a pot over medium heat. Break them up and add turkey, salt and herbs. Continue cooking until turkey is ready.

5. Cook tortillas on a bare stove burner on low heat.

6. Add meat onto tlacho and serve.

To extend their agriculture, the Aztecs built artificial garden islands called **chinampas** on the freshwater lakebeds in the Valley of Mexico. Xochimilco, which already used small scale chinampas farming, was defeated and brought into the Aztec Triple Alliance. These artificial islands, 30 x 2.5 metres, were created by interweaving reeds with stakes beneath the lake's surface which created underwater fences. Layers of soil and aquatic vegetation would be piled upon these fences until the top layer of soil broke the surface of the water. Often trees such as willow and cypress were planted in the corners to secure the chinampas. The soil from the bottom of the lake was rich in nutrients which fertilised the chinampas. These raised beds were very fertile, with up to 7 harvests a year of crops such as maize, beans, squash, amaranth, tomatoes, chili peppers and flowers.

A drainage system was also developed which allowed for the flow of water and sediments. The chinampas were separated by channels wide enough for a canoe to pass. Xochimilico still has the 6,000 acres of canals, with some of the chinampas still growing the traditional crops as well as beets, celery, Brussels sprouts and various herbs.

Modern painting of chinampas

MEXICO currently suffers from food insecurity for at least 10% of the population because of access, not availability. The pandemic hit Mexico hard in 2020 and pushed up to ten million people into extreme poverty. A group of students created the Farmlink Project, a non-profit organisation to alleviate food insecurity among the impoverished, who were left to fend for themselves by the government. The project finds inefficiencies in the food distribution system that leads to food waste. By implementing measures to prevent waste and receiving donations, the Project helps transfer that food directly to impoverished communities through food banks.

CENTRAL AMERICA has been suffering a devastating drought for at least five years which has impoverished many indigenous people. It is believed that the drought destroyed between 50-90% of the region's agricultural production in 2020, at a time when the

Covid pandemic affected agricultural day labourers. Many subsistence farms also suffered damage from hurricanes Eta and Iota that year, with huge losses of cash crops, staples and coffee farmland. Hunger in Central America increased from 2.2 million people in 2018 to nearly 8 million in 2021.

The Dry Corridor consists of Guatemala, Honduras, El Salvador and Nicaragua. Maize and beans, food staples in the region, were most affected by the prolonged drought.

In 2022, Guatemala was the highest food risk country in Latin America, followed by Honduras, El Salvador and Nicaragua. Basic grain reserves have been depleted in rural areas since the Ukraine war led to complications in the global supply chain and increases in the cost of food and fuel.

Food insecurity in Central America is also one of the driving factors leading to the huge wave of illegal migrants trying to enter the U.S.A. The WFP announced a six month assistance package of $47 million to assist 2.6 million of the most adversely affected people across the four countries in 2021.

SOUTH AMERICA

South America has three areas which are believed to have independently invented agriculture:

- The Andes highlands
- The Peruvian coastline, desert cultures
- Amazonian and Colombian jungles

THE ANDES HIGHLANDS

The highland Andes region was one of the earliest areas to transition from hunting/gathering to agriculture. For thousands of years before the Inca Empire, the Andean people developed an agricultural technique called *Waru Waru* which was based on modification of the soil surface to store water and use it for irrigation. They combined raised garden beds with irrigation channels to increase crop yields and prevent soil erosion by floods, which allowed them to grow crops at 12,500 feet above sea level. The system was also known as *suqakollos.* As many as 250,000 acres around Lake Titicaca show traces of early Waru Waru technology.

In 1981, Clark Erickson of the University of Illinois, persuaded farmers to rebuild some of the raised beds to grow potatoes, quinoa and canihua. Locals from an Aymara community on the shore of Lake Titicaca claimed that the Waru Waru helped them to make the land productive again, and even created its own microclimate. Other scientists believe that the Waru Waru technology could be used in conjunction with other ancient technologies, such as terraces known as *andenes,* and interconnected irrigation lakes called *qochas.*

REFERENCE: Lloyd, Ellen, “Waru waru- Ancient Andean Irrigation System brought Back to Life,” “Ancient Pages,” April 28, 2018

Cumbemayo, which is located about 12 miles (19km) southwest of the Peruvian city of Cajamarca, is known for its ancient aqueduct which predates the Inca Empire. Believed to have been constructed around 1500 BCE, it carried water from the hills and redirected it to cultivation fields and a reservoir. The sophisticated aqueduct is five miles long and carved out of volcanic rock at an altitude of 11,000 feet.

Archeologists are unsure who built the aqueduct and carved into the stone, but it is possible that the Chavin culture was responsible. It was once believed to be the oldest structure in South America.

The potato was domesticated between 8000 and 5000 BCE, along with beans, tomatoes, peanuts, cacao, llamas, alpacas and guinea pigs. From about 7000 BCE, maize was imported from

Mesoamerica and genetically modified to become domestic maize. Cotton was domesticated in Peru by 4200 BCE.

Cumbemayo aqueduct, credit Hidden Inca Tours.

Potatoes are indigenous to South America and may have been cultivated up to 10,000 years ago, but the oldest tuber remains have been found at Ancon in Peru, dating to 2500 BCE. The shape of ceramic vessels in the Altiplano was also influenced by the potato, indicating its importance. Indigenous Andeans prepared their potatoes by boiling, baking, mashing or stewing. One of their dishes, *papas secas*, involved boiling, peeling and chopping potatoes which were then fermented to create *toqosh*. Finally, they were ground to a pulp, soaked and filtered into a starch known as *almidon de papa*.

Moche potato ceramic, credit Pattych, CC BY-SA 3.0

Chuño is a freeze dried potato product made by the Aymara and Quechua which predated the Inca Empire. The five-day process starts by exposing a frost-resistant variety of potatoes to freezing night temperatures of the Altiplano and subsequently exposing them to the intense sunlight. The dessicated potatoes were able to be stored for years at a time and made into a flour which is still used in traditional Bolivian dishes.

Capsicum and peppers originated in the highlands of Bolivia where they have been eaten since around 7500 BCE. Members of the pepper group such as chiles, red peppers, jalapenos and chipotles (smoked japapenos) have no connection with the Asian black pepper group. They spread to Central America between 5200–3400 BCE and were used in Olmec, Maya and Aztec cuisines.

The oldest **cotton** fabric has been found in Huaca Prieta in Peru, dated to about 6000 BCE, so it must have been domesticated earlier. Other cotton balls were discovered in a cave in the Tehuacán Valley, Mexico dated to about 5500 BCE. Seeds dating to 2500 BCE have been discovered in Peru, and by that time cotton was being grown and woven in Mexico and Arizona.

The **Incan Empire** began around mid-1400 CE with its capital in Cuzco, high in the Andes. Machu Picchu, the most famous Incan site, contains a complex aqueduct system. They cut the canals out of stone, lined them with rock and filled the joints with clay to reduce water loss to seepage. The aqueducts provided enough water for sixteen fountains for domestic use. The town of Moray had three depressions of concentric terraced circles, with vertical channels dropping water from one level of the terrace to the next.

Tipon near Cuzco had aqueducts set atop walls that may have been built by the earlier Wari culture. The Inca reinforced the Wari wall by using andesite, as well as improving three existing canals. Of the thirteen terraces, eleven are irrigated by water from the monumental Tipon spring. The pictured monumental fountain was restored in 1999.

Tipon monumental fountain, credit Khan (IT) CC BY-SA 4.0

Farming and Food

Agriculture was often terraced because of the mountainous terrain. Stone and wooden tools included the hoe and foot plough which consisted of a wooden pointed pole that was pushed into the ground using one's foot on a horizontal bar. Farmers worked in small teams of about eight, with the men hoeing the ground so that the women following behind could break clods and sow seeds. Older children tended the llama and alpaca herds which provided wool, leather, meat and transportation. The Incas practised crop rotation and fertilisation using llama dung or guano.

Crops grown included potato, maize, cocoa, sweet potatoes, ulluco, chiles, tomatoes, squash, cucumber, gourd, avocado and other relatively unknown vegetables such as cherimoya, lucuma and guayabo,

RECIPE: Olluco con Carne is a dish which originated in Inca times or even before. Olluco is a native Andean tuber which has been cultivated for at least 4,000 years. It is a cousin to the potato but less starchy with a soft texture which is unsuitable for fries. However it is perfect for stews. In this adapted recipe, it can be substituted with potato.

Ingredients:

- 1 lb sirloin steak cut into strips (or traditional charqui soaked overnight)
- 2.2 lbs ollucos julienned into thin strips
- 2 large sweet potatoes peeled and boiled
- 1 lb long-grain boiled rice
- 1 onion diced
- 1 fresh aji Amarillo chili pepper chopped into strips
- 1 tbsp aji panca chili pepper paste
- ¼ tsp cumin, 1 tsp oregano
- 2 tbsp vegetable or olive oil

Method: 1. Lightly fry onion, garlic, pepper paste and oregano in oil for 3 minutes and then add the aji Amarillo chili pepper.

2. Add beef strips and fry until golden.

3. Add julienned olluco with a pinch of salt. Mix. And fry for 20 minutes.

4. Serve with rice and sweet potatoes as sides.

Image and recipe source: eat peru
https://www.eatperu.com/olluco-con-carne/

COASTAL REGIONS OF PERU

Caral is the oldest city in South America and contemporaneous with the pyramids of Egypt (2600-2000 BCE.) The inhabitants built monumental architecture including large earthwork platforms and sunken circular plazas, but no pottery has yet been discovered, making it part of the pre-ceramic Archaic period. The city was built 23 km from the coast and 350 metres above sea level in an arid environment. Irrigation canals were used to grow crops such as squash, beans, guava, pacay, sweet potato, avocado and maize. Protein came mainly from seafood such as clams, mussels, anchovies, sardines and to a lesser extent, sea mammals. Most of this seafood came from Aspero, an early fishing town (3700-2500 BCE,) which supported an ancient Peruvian culture based upon fishing, shellfish and hunting sea mammals, rather than agriculture.

The **Chavin** culture developed in the Andean highlands of Peru from 900-200 BCE and spread to the arid coastal regions. The main centre was Chavin de Huantar in the highlands, where drainage and irrigation canals were built to cultivate numerous crops such as potatoes, quinoa and maize. They domesticated llamas and alpacas for meat, fibre and as pack animals. Their pottery is famous for its lewd, sexual images.

The **Paracas** (700-200 BCE) and Nazca (100 BCE-750 CE) cultures of coastal Peru carried out extensive hydraulic works, such as irrigation works for growing crops in the arid region. The Nazca built an extensive system of subterranean aqueducts, channels and reservoirs called **puquios** to distribute water to farmland and villages. Along the course of many underground channels are spiralling well-constructions known as ojos which helped to funnel wind into the underground canals. Nazca subsistence was mainly based on agriculture, including crops like maize, squash, beans, sweet potatoes, peanuts and manioc. Non-food crops included cotton, cocoa, cactus and gourds. Cocoa leaves were chewed and worked as a stimulant, hunger suppressant and to combat fatigue. The San Pedro cactus provided hallucinogenic properties for ceremonial occasions. Some seafood, llamas and guinea pigs were also eaten.

It is believed that the puquios system was built upon similar principles to the qanats of Iran, although there was no connection. Numerous puquios date from about 500 CE, but many of the elaborate ones were commissioned by the Spanish after the conquest of Peru in the 16th century.

AMAZONIA, the world's largest forest and river basin covers 2.6 million square miles (6.7 mill square kms) and holds an extraordinary array of life. Once considered a pristine wilderness since time immemorial, archeologists have quite recently discovered evidence of large complex societies who domesticated plants and animals. At least 83 native Amazonian species were domesticated, such as sweet potato, cacao, tobacco, pineapple, cassava, passionfruit and hot peppers.

The dark soils were rich and possibly supported communities occupying as much as 0.1% of the Amazonian rainforest which straddles numerous countries. These dark soils, which are rich in nutrients and carbon from human waste, fires, mulching and composting on farms, first appeared in parts of Amazonia 6,000 years ago. They increased rapidly in size and number about 2,500 years ago, but unfortunately, in the decades after the Spanish conquest, most of these societies disappeared.

One large group, the Arawak, originated in western Amazonia and expanded across the vast river areas which were associated with the development of farming villages and agricultural landscapes. They also spread to Caribbean islands such as Trinidad, Jamaica and Grenada.

In 2020, researchers from the University of Bern, led by Umberto Lombardo, used remote sensing to investigate the savannah known as Llanos de Moxos in north east Bolivia and detected more than 6,600 areas that had once been forested. Samples of sediments yielded phytoliths of cassava dated to 10,350 years ago, squash- 10,250 BP and maize 6850 BP, making it one of the earliest agricultural sites in South America.

"Early Agricultural Hotspot found in Amazonia," "Archeology," April 9, 2020

https://www.archaeology.org/news/8583-200409-bolivia-amazonia-agriculture

Crops from Amazonia include:

- **Passionfruit** originated in the Amazonian region of Brazil or Colombia. The vine was cultivated by the Inca and Aztecs.

- **Pineapples** originated from the riverine areas of southern Brazil and Paraguay. Archeological evidence of cultivation has been found and dated from Peru (3200-2800 BP) and Mexico (2200-2800 BP.) The Mayans and Aztecs also cultivated it.
- **Cashew nuts** are native to Amazonian Brazil and Venezuela. The plant comes in two parts, the stem, known as the cashew-apple and the nut, the true fruit.
- **Peanuts** originated in Brazil or Peru. For thousands of years, tribes were creating jars shaped like peanuts and decorated with them. **Brazil nuts** are one of the largest and longest-live trees in the Amazonian forest.
- **Acai** palms are native to eastern Amazonia, mainly in swamps and floodplains. The purple fruit has been a staple of Amazonian societies for centuries, if not longer.
- **Cacao** has been cultivated in Amazonia since 5,300 BP. It was introduced to Mexico by the Olmecs more than 4,000 years ago. Chocolate was made from the seed.
- **Guaraná** is native to Maués in Amazonian Brazil and has been used for millennia by the Sateré-Mawé indigenous people as a stimulant due to its high caffeine content.
- **Guava** is native to tropical South and Central America. Earliest evidence of cultivation comes from tropical areas of Peru as early as 4500 BP.
- Less known fruits include Camu Camu, (rumberry) bacaba palm, cupuacu, aguaje, (moriche palm fruit) cocona, sacha inchi, and pacay which tastes similar to vanilla icecream.
- **Manioc,** also known as cassava, has tuberous roots from which flour breads and tapioca are derived. It provided vital carbohydrates for rainforest dwellers but had to be properly prepared to remove the cyanide producing sugar derivatives.

Pacay fruit

Maracuya, or passionfruit

Images credit https://www.rainforestcruises.com/guides/amazon-rainforest-fruits

SOUTH AMERICAN food insecurity had been improving since the U.N. adopted a Millennium Development Goal in 2000 to cut hunger in half by 2015. Bolivia reduced its hunger scale from 38% to 15.9%, while Peru made even greater strides by reducing its percentage of hunger from 31.6 to 7.5 in 2015.

However, from 2019-20, economic collapse in Venezuela, rampant inflation and the Covid-19 pandemic saw malnutrition rise to its worst levels in decades, with 13.8 million people suffering food insecurity in South America according to a United Nations report. In 2020, a group of national universities, ENVOVI, conducted a survey that concluded 74% of Venezuelan households suffer from extreme poverty and food insecurity. WFP set up the Venezuela Humanitarian Response Place with Humanitarian Needs Overview 2020 to help the most vulnerable. Cuatro por Venezuela Foundation focuses on a nutrition program to provide food staples to orphanages, nursing homes, schools, hospitals and homeless organisations.

The Ukraine war in 2022 has seen the price of food skyrocket across Latin America, especially in Argentina where wheat prices rose by 90%.

BOLIVIA remains the second most food insecure country in South America, after Venezuela.

According to Marianne Fay, the World Bank director for Bolivia, "High inflation affecting food and energy prices, the impact of COVID-19 and climate change are having an impact on rural producer households, pushing more people into extreme poverty and hindering the fight against hunger and global malnutrition. For this reason, in Bolivia and several other countries of the region, the World Bank supports investment in climate-resilient agriculture, the promotion of sustainable food production and the transformation of food systems Agriculture is a key sector for the Bolivian economy, with significant potential to reduce rural poverty and dependence on extractive sectors," she said.

In August 2022, the World Bank approved a USD300 million loan to benefit 130,000 farmers in Bolivia.

World Bank Press Release, "The World Bank Will Support Bolivian Producers to Increase Food Security and Market Access," August 10, 2022

https://www.worldbank.org/en/news/press-release/2022/08/10/the-world-bank-will-support-bolivian-producers-to-increase-food-security-and-market-access

NORTH AMERICA

At least three agricultural complexes originated north of the Rio Grande in the southwest USA; the Upper Sonoran Complex, the Lower Sonoran Complex and the Eastern Agricultural Complex. Corn appears to be the first crop cultivated in the southwest, with the Bat cave in New Mexico yielding remains of the 3,500-year-old cultigen. Squash was also present at this site. Beans appeared about 500 CE.

The Upper Sonoran complex of New Mexico was high altitude farming above 4,500 feet, where crops like corn, squash, gourd and beans were found in areas with rainfall greater than 200 mm (8 inches) annually. The Lower Sonoran complex, with less annual precipitation, included squash, corn, cotton, and beans (lima, scarlet runner and jack bean.) These crops were integrated into the diets of Archaic cultures which were characterised by high mobility, no pottery and extensive plant cultivation, including grain harvesting.

Around 1,700 years ago, the Archaic cultures became more sedentary and food production was more widespread. From 1100 CE, The Ancestral Pueblo, Hohokam and Morgollon communities were using sophisticated agricultural techniques, including irrigation, hillside contour terraces and bordered gardens. At Snaketown in Arizona, a complex canal system with many 2 metre (6.5 ft) deep and 3 metre (10 feet) wide canals was constructed. Hundreds of kilometres of canals have also been uncovered in the nearby Phoenix area.

The third agricultural regime was found in the region between the Mississippi River and Appalachian Mountains, which includes Illinois, Kentucky and Tennessee. Plants grown in the Eastern Agricultural Complex included sunflower, squash, amaranth, maygrass, sumpweed, little barley and knotweed. Fish, shellfish, deer, acorns, walnuts and hickory nuts were also consumed. The earliest domesticated plant in the region, squash, was uncovered from sites in Missouri, Illinois, Kentucky and Maine dating from 8,000 to 5,000 years ago. Sunflower was first domesticated in the east, with small sunflower fruits discovered from the Koster site from 9000 BP. By 5000 BP, larger domesticated sunflower fruits were reported from the Hayes site in Tennessee.

- Chenopods, also called goosefoot, were domesticated at least 4,500 years ago in Kentucky and Illinois. This family of plants includes spinach and quinoa.
- Sumpweed fruits were harvested in Illinois 7,000 years ago, and by 5500 BP a domesticated sumpweed was being grown.
- Corn was grown in the central Mississippi valley about 2100 BP, and five hundred years later was grown as far north as Mississippi.
- Pecans were indigenous and closely related to the history of native Americans.
- Charred seeds found in the Utah desert show that tobacco has been used for over 10,000 years for sacred and medicinal purposes. Prior to this discovery, the earliest known evidence of tobacco was nicotine found in Alabama smoking pipes dating back 3,300 years. The crop was used across the Americas.

- Turkeys are indigenous to North and Central America. The wild turkey, *Meleagris gallopavo* was domesticated in parts of the USA over 2,000 years ago, while the ocellated turkey is native to the Yucatan Peninsula.
- Pumpkins have been domesticated as early as 7,000-5,500 BCE in the USA and north east Mexico.

Native tribes living from British Columbia to California managed their habitats and plants. The Owens Valley Paiute irrigated the grasses they used for subsistence, while other groups controlled burning to increase acorn production by burning the oak trees. Often they planted tobacco in the burned areas.

REFERENCE: Britannica: Origins of agriculture, North America

Mixed cropping, whereby different crops are planted together, was practised in Mesoamerica 5,000 years ago and eventually spread to North America. "The Three Sisters" included corn, beans and squash. The corn was planted first which provided a stalk for the beans, whereas the squash plant was grown on the ground, shaded by the beans and corn and keeping the weeds from the other two plants. Corn would deplete the nitrogen from the soil, while beans would replace it into the soil.

The Iroquois Legend of the Three Sisters

"The term "Three Sisters" emerged from the Iroquois creation myth. It was said that the earth began when "Sky Woman" who lived in the upper world peered through a hole in the sky and fell through to an endless sea. The animals saw her coming, so they took the soil from the bottom of the sea and spread it onto the back of a giant turtle to provide a safe place for her to land. This "Turtle Island" is now what we call North America.

"Sky woman had become pregnant before she fell. When she landed, she gave birth to a daughter. When the daughter grew into a young woman, she also became pregnant (by the West wind). She died while giving birth to twin boys. Sky

> Woman buried her daughter in the "new earth." From her grave grew three sacred plants—corn, beans, and squash. These plants provided food for her sons, and later, for all of humanity. These special gifts ensured the survival of the Iroquois people."

Erney, Diana, "Long live the Three Sisters," "Organic Gardening." November, 1996

Squash is a native American plant named from the Narragansett language. They are plants in the Cucurbita family, made up of about 15 species including pumpkin, and are divided into summer and winter squashes. Many squashes originated in Mesoamerica at least 7,000 years ago.

- Summer squashes : crookneck, scallop, straightneck, marrow, zucchini
- Winter squashes, marrow, pumpkin, acorn, butternut

RECIPE: Cherokee style Succotash using mainly native ingredients.

- 2 cups of Lima, kidney or black beans, cooked in advance
- 2 cups of corn, cooked
- 1 cup of julienned bell peppers, 3 colours if available
- 1 large onion, chopped
- 3 garlic cloves, finely chopped
- 1 fresh tomato, finely diced
- 2 or 3 tbsp of olive oil
- 2 sprigs of fresh thyme & 2 stalks of spring onions, finely chopped
- 1/2 tsp of black pepper and 1 pinch of chili powder (optional)
- Salt, to taste

Method: 1. Fry garlic, onion, peppers, tomato, corn and beans in that order.

2. Incorporate herbs, spices, fry on lower heat until cooked. Serve warm.

https://culinaremundi.com/recipe/cherokee-style-succotash/

ORIGINS OF FRUIT AND VEGETABLES

- Apple- Europe, West Asia
- Apricot- Central Asia, China, Europe
- Asparagus- Europe, Africa
- Avocado- Mexico
- Banana- New Guinea
- Beans- Mexico
- Blueberry- North America, Europe, Asia
- Capsicum- Bolivia
- Carrot- Europe, Persia
- Cherry- Asia
- Chickpeas- Middle East
- Corn- Mexico
- Cucumber- India
- Eggplant- India
- Gooseberry- China
- Grape- Fertile Crescent, Mediterranean
- Guava- Mexico and Amazonia
- Lemon- Himalayas
- Lentils- Middle East
- Lime- South East Asia
- Mandarin- China
- Mulberry- China
- Onion- Central Asia
- Orange- Himalayas
- Plum- Iran and China
- Pumpkin- Mexico, SW USA
- Passionfruit- Amazon/Colombia
- Pineapple- Brazil
- Potato- Peru/Bolivia
- Shallot- tropical Asia
- Strawberry- Europe
- Sweet potato- Central America
- Tomato- Mexico

- Watermelon- Africa
- Zucchini- Mexico

ORIGINS OF GRAINS

- Wheat- Fertile Crescent
- Millet- Africa, Middle East, India
- Barley- Fertile Crescent
- Rye- Fertile Crescent
- Corn- Mexico
- Spelt- Iran
- Rice- China and India

ORIGINS OF HERBS, SPICES and NUTS

- Acai- Amazonia
- Anise- Mediterranean
- Asafoetida- Afghanistan
- Black pepper- India
- Brazil nuts- Amazonia
- Cardamon- India
- Cashew nuts- Amazonia
- Chili- Central America
- Chipotle- Mesoamerica
- Cinnamon- Sri Lanka
- Cloves- Indonesia
- Coriander- Mediterranean and Middle East
- Cumin- Mediterranean
- Fennel- Mediterranean, Eurasia
- Galangal- South East Asia
- Garlic- Central Asia
- Ginger- China
- Harissa- North Africa
- Macadamia nuts- Australia
- Mace- Indonesia
- Mint- Europe, Asia and the Americas

- Nutmeg- Indonesia
- Oregano- Mediterranean
- Paprika, ground peppers-Mexico
- Peanut- Amazonia
- Sage- Europe and western Asia
- Thyme- Mediterranean
- Turmeric- India
- Walnut- Europe and Asia

AFTERWORD

"The World Food Programme (WFP), run under the auspices of the United Nations, has recently designated hunger as the new normal. Its recently published Global Operational Response Plan 2022 warns that the world is confronting the most significant food crisis in modern history, which might make a staggering 345.2 million people food insecure in 2023. This number will be more than double the number of people who face food insecurity in 2020.

"This 'new normal' has its origin in a triple-C crisis: the ongoing conflict in Ukraine which has pushed the global food and fuel prices soaring, the low agricultural yields (and reduced purchasing power) resulting from climate change, and the aftershocks of supply chain disruptions caused by the Covid-19 pandemic.

"The WFP has highlighted all three factors – as well as an increase in the cost of obtaining bank loans, deterioration of macro-economic indicators in most countries across the world, and unsustainable amount of debt which is affecting not just advanced economies such as that of the United States but also developing economies such as Pakistan's – and cautioned that the level of food insecurity could escalate in several countries, including Sri Lanka, Ghana, Pakistan, Tunisia, Egypt, Kenya, and Laos."

https://www.thenews.com.pk/print/1051214-food-insecurity-in-pakistan

As the world increasingly suffers from food insecurity brought about by war, pestilence and famine, many people are eager to learn about the foods our distant ancestors grew, produced and consumed. Ancient practices like fermentation and preservation are once again coming into focus as people are eager to study their ancestral roots. Artisan breads, cheeses, wines, beer and other ancient fermented products are gaining in popularity as consumers eschew factory produced foods which are laced with artificial chemicals.

Now more than ever, it is time to learn how to grow, process and cook ancestral foods, utilising practices such as fermentation and traditional farming methods. Our future as a species depends on it!

JACK THE RIPPER'S NEW TESTAMENT
Occultism and Bible Mania in 1888
By Nigel Graddon

This book offers evidence, for the first time, that those responsible for the Whitechapel murders were members of a hit team associated with a centuries-old European occult confederacy dedicated to human sacrifice. This was corroborated in the private papers of a Monsignor who carried out intelligence work for Pope Pius X in the run-up to the outbreak of global conflict in 1914. The priest told of the existence of a Vatican-based cabal of assassins formed by the infamous Borgias that is in alliance with a Teuton occult group formed in the 9th century. It was from within this unholy alliance that assassins travelled to London to carry out the Ripper murders to "solve a sticky problem for the British Royal Family"

302 Pages. 6x9 Paperback. Illustrated. $19.95. Code: JRNT

SECRETS OF THE HOLY LANCE
The Spear of Destiny in History & Legend
by Jerry E. Smith

Secrets of the Holy Lance traces the Spear from its possession by Constantine, Rome's first Christian Caesar, to Charlemagne's claim that with it he ruled the Holy Roman Empire by Divine Right, and on through two thousand years of kings and emperors, until it came within Hitler's grasp—and beyond! Did it rest for a while in Antarctic ice? Is it now hidden in Europe, awaiting the next person to claim its awesome power? Neither debunking nor worshiping, *Secrets of the Holy Lance* seeks to pierce the veil of myth and mystery around the Spear.

312 PAGES. 6x9 PAPERBACK. ILLUSTRATED. $16.95. CODE: SOHL

THE CRYSTAL SKULLS
Astonishing Portals to Man's Past
by David Hatcher Childress and Stephen S. Mehler

Childress introduces the technology and lore of crystals, and then plunges into the turbulent times of the Mexican Revolution form the backdrop for the rollicking adventures of Ambrose Bierce, the renowned journalist who went missing in the jungles in 1913, and F.A. Mitchell-Hedges, the notorious adventurer who emerged from the jungles with the most famous of the crystal skulls. Mehler shares his extensive knowledge of and experience with crystal skulls. Having been involved in the field since the 1980s, he has personally examined many of the most influential skulls, and has worked with the leaders in crystal skull research. Color section.

294 pages. 6x9 Paperback. Illustrated. $18.95. Code: CRSK

THE LAND OF OSIRIS
An Introduction to Khemitology
by Stephen S. Mehler

Was there an advanced prehistoric civilization in ancient Egypt? Were they the people who built the great pyramids and carved the Great Sphinx? Did the pyramids serve as energy devices and not as tombs for kings? Chapters include: Egyptology and Its Paradigms; Khemitology—New Paradigms; Asgat Nefer—The Harmony of Water; Khemit and the Myth of Atlantis; The Extraterrestrial Question; more. Color section.

272 PAGES. 6x9 PAPERBACK. ILLUSTRATED . $18.95. CODE: LOOS

VIMANA:
Flying Machines of the Ancients
by David Hatcher Childress

According to early Sanskrit texts the ancients had several types of airships called vimanas. Like aircraft of today, vimanas were used to fly through the air from city to city; to conduct aerial surveys of uncharted lands; and as delivery vehicles for awesome weapons. David Hatcher Childress, popular *Lost Cities* author, takes us on an astounding investigation into tales of ancient flying machines. In his new book, packed with photos and diagrams, he consults ancient texts and modern stories and presents astonishing evidence that aircraft, similar to the ones we use today, were used thousands of years ago in India, Sumeria, China and other countries. Includes a 24-page color section.

408 Pages. 6x9 Paperback. Illustrated. $22.95. Code: VMA

THE LOST WORLD OF CHAM
The Trans-Pacific Voyages of the Champa
By David Hatcher Childress

The mysterious Cham, or Champa, peoples of Southeast Asia formed a megalith-building, seagoing empire that extended into Indonesia, Tonga, and beyond—a transoceanic power that reached Mexico and South America. The Champa maintained many ports in what is today Vietnam, Cambodia, and Indonesia and their ships plied the Indian Ocean and the Pacific, bringing Chinese, African and Indian traders to far off lands, including Olmec ports on the Pacific Coast of Central America. opics include: Cham and Khem: Egyptian Influence on Cham; The Search for Metals; The Basalt City of Nan Madol; Elephants and Buddhists in North America; The Olmecs; The Cham in Colombia; tons more. 24-page color section.

328 Pages. 6x9 Paperback. Illustrated. $22.00 Code: LPWC

OTTO RAHN & THE QUEST FOR THE HOLY GRAIL
The Amazing Life of the Real "Indiana Jones"
By Nigel Graddon

Otto Rahn, a Hessian language scholar, is said to have found runic Grail tablets in the Pyrenean grottoes, unearthed as a result of his work in decoding the hidden messages within the Grail masterwork *Parsifal*. The fabulous artifacts identified by Rahn were believed by Himmler to include the Grail Cup, the Spear of Destiny, the Tablets of Moses, the Ark of the Covenant, the Sword and Harp of David, the Sacred Candelabra and the Golden Urn of Manna. Some believe that Rahn was a Nazi guru who wielded immense influence within the Hitler regime, persuading them that the Grail was the Sacred Book of the Aryans, which, once obtained, would justify their extreme political theories.

450 pages. 6x9 Paperback. Illustrated. Index. $18.95. Code: ORQG

THE LANDING LIGHTS OF MAGONIA
UFOs, Aliens and the Fairy Kingdom
By Nigel Graddon

British UFO researcher Graddon takes us to that magical land of Magonia—the land of the Fairies—a place from which some people return while others go and never come back. Graddon on fairies, the wee folk, elves, fairy pathways, Welsh folklore, the Tuatha de Dannan, UFO occupants, the Little Blue Man of Studham, the implications of Mars, psychic connections with UFOs and fairies. He also recounts many of the strange tales of fairies, UFOs and Magonia. Chapters include: The Little Blue Man of Studham; The Wee Folk; UFOlk; What the Folk; Grimm Tales; The Welsh Triangle; The Implicate Order; Mars—an Atlantean Outpost; Psi-Fi; High Spirits; "Once Upon a Time…"; more.

270 Pages. 6x9 Paperback. Illustrated. $19.95. Code: LLOM

ADVENTURES OF A HASHISH SMUGGLER
by Henri de Monfreid

The son of a French artist who knew Paul Gaugin as a child, de Monfreid sought his fortune by becoming a collector and merchant of the fabled Persian Gulf pearls. He was then drawn into the shadowy world of arms trading, slavery, smuggling and drugs. Infamous as well as famous, his name is inextricably linked to the Red Sea and the raffish ports between Suez and Aden in the early years of the twentieth century. De Monfreid (1879 to 1974) had a long life of many adventures around the Horn of Africa where he dodged pirates as well as the authorities.

284 Pages. 6x9 Paperback. $16.95. Illustrated. Code AHS

TECHNOLOGY OF THE GODS
The Incredible Sciences of the Ancients
by David Hatcher Childress

Childress looks at the technology that was allegedly used in Atlantis and the theory that the Great Pyramid of Egypt was originally a gigantic power station. He examines tales of ancient flight and the technology that it involved; how the ancients used electricity; megalithic building techniques; the use of crystal lenses and the fire from the gods; evidence of various high tech weapons in the past, including atomic weapons; ancient metallurgy and heavy machinery; the role of modern inventors such as Nikola Tesla in bringing ancient technology back into modern use; impossible artifacts; and more.

356 pages. 6x9 Paperback. Illustrated. $16.95. code: TGOD

THE ANTI-GRAVITY HANDBOOK
edited by David Hatcher Childress

The new expanded compilation of material on Anti-Gravity, Free Energy, Flying Saucer Propulsion, UFOs, Suppressed Technology, NASA Cover-ups and more. Highly illustrated with patents, technical illustrations and photos. This revised and expanded edition has more material, including photos of Area 51, Nevada, the government's secret testing facility. This classic on weird science is back in a new format!

230 pages. 7x10 paperback. Illustrated. $16.95. code: AGH

ANTI–GRAVITY & THE WORLD GRID

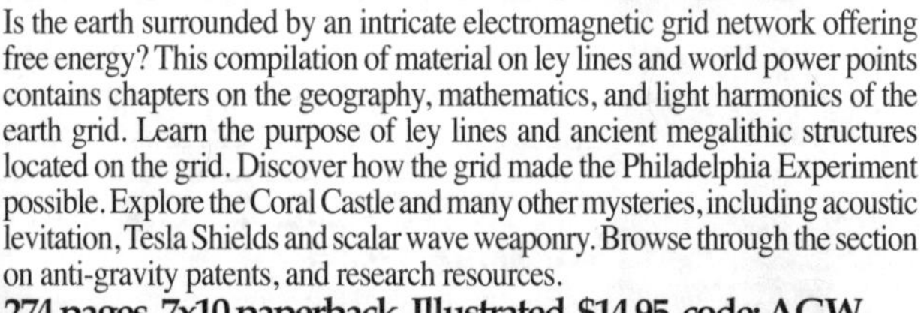

Is the earth surrounded by an intricate electromagnetic grid network offering free energy? This compilation of material on ley lines and world power points contains chapters on the geography, mathematics, and light harmonics of the earth grid. Learn the purpose of ley lines and ancient megalithic structures located on the grid. Discover how the grid made the Philadelphia Experiment possible. Explore the Coral Castle and many other mysteries, including acoustic levitation, Tesla Shields and scalar wave weaponry. Browse through the section on anti-gravity patents, and research resources.

274 pages. 7x10 paperback. Illustrated. $14.95. code: AGW

ANTI–GRAVITY & THE UNIFIED FIELD
edited by David Hatcher Childress

Is Einstein's Unified Field Theory the answer to all of our energy problems? Explored in this compilation of material is how gravity, electricity and magnetism manifest from a unified field around us. Why artificial gravity is possible; secrets of UFO propulsion; free energy; Nikola Tesla and anti-gravity airships of the 20s and 30s; flying saucers as superconducting whirls of plasma; anti-mass generators; vortex propulsion; suppressed technology; government cover-ups; gravitational pulse drive; spacecraft & more.

240 pages. 7x10 Paperback. Illustrated. $14.95. Code: AGU

THE TIME TRAVEL HANDBOOK
A Manual of Practical Teleportation & Time Travel
edited by David Hatcher Childress

The Time Travel Handbook takes the reader beyond the government experiments and deep into the uncharted territory of early time travellers such as Nikola Tesla and Guglielmo Marconi and their alleged time travel experiments, as well as the Wilson Brothers of EMI and their connection to the Philadelphia Experiment—the U.S. Navy's forays into invisibility, time travel, and teleportation. Childress looks into the claims of time travelling individuals, and investigates the unusual claim that the pyramids on Mars were built in the future and sent back in time. A highly visual, large format book, with patents, photos and schematics. Be the first on your block to build your own time travel device!

316 pages. 7x10 Paperback. Illustrated. $16.95. code: TTH

ANCIENT ALIENS ON THE MOON
By Mike Bara

What did NASA find in their explorations of the solar system that they may have kept from the general public? How ancient really are these ruins on the Moon? Using official NASA and Russian photos of the Moon, Bara looks at vast cityscapes and domes in the Sinus Medii region as well as glass domes in the Crisium region. Bara also takes a detailed look at the mission of Apollo 17 and the case that this was a salvage mission, primarily concerned with investigating an opening into a massive hexagonal ruin near the landing site. Chapters include: The History of Lunar Anomalies; The Early 20th Century; Sinus Medii; To the Moon Alice!; Mare Crisium; Yes, Virginia, We Really Went to the Moon; Apollo 17; more. Tons of photos of the Moon examined for possible structures and other anomalies.
248 Pages. 6x9 Paperback. Illustrated.. $19.95. Code: AAOM

ANCIENT ALIENS ON MARS
By Mike Bara

Bara brings us this lavishly illustrated volume on alien structures on Mars. Was there once a vast, technologically advanced civilization on Mars, and did it leave evidence of its existence behind for humans to find eons later? Did these advanced extraterrestrial visitors vanish in a solar system wide cataclysm of their own making, only to make their way to Earth and start anew? Was Mars once as lush and green as the Earth, and teeming with life? Chapters include: War of the Worlds; The Mars Tidal Model; The Death of Mars; Cydonia and the Face on Mars; The Monuments of Mars; The Search for Life on Mars; The True Colors of Mars and The Pathfinder Sphinx; more. Color section.
252 Pages. 6x9 Paperback. Illustrated. $19.95. Code: AMAR

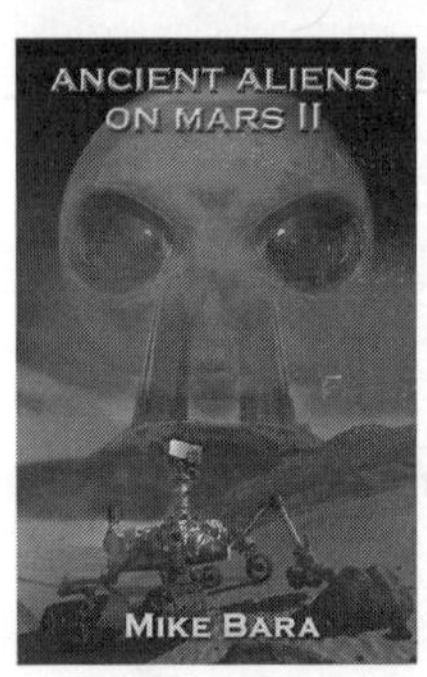

ANCIENT ALIENS ON MARS II
By Mike Bara

Using data acquired from sophisticated new scientific instruments like the Mars Odyssey THEMIS infrared imager, Bara shows that the region of Cydonia overlays a vast underground city full of enormous structures and devices that may still be operating. He peels back the layers of mystery to show images of tunnel systems, temples and ruins, and exposes the sophisticated NASA conspiracy designed to hide them. Bara also tackles the enigma of Mars' hollowed out moon Phobos, and exposes evidence that it is artificial. Long-held myths about Mars, including claims that it is protected by a sophisticated UFO defense system, are examined. Data from the Mars rovers Spirit, Opportunity and Curiosity are examined; everything from fossilized plants to mechanical debris is exposed in images taken directly from NASA's own archives.
294 Pages. 6x9 Paperback. Illustrated. $19.95. Code: AAM2

ANCIENT TECHNOLOGY IN PERU & BOLIVIA
By David Hatcher Childress

Childress speculates on the existence of a sunken city in Lake Titicaca and reveals new evidence that the Sumerians may have arrived in South America 4,000 years ago. He demonstrates that the use of "keystone cuts" with metal clamps poured into them to secure megalithic construction was an advanced technology used all over the world, from the Andes to Egypt, Greece and Southeast Asia. He maintains that only power tools could have made the intricate articulation and drill holes found in extremely hard granite and basalt blocks in Bolivia and Peru, and that the megalith builders had to have had advanced methods for moving and stacking gigantic blocks of stone, some weighing over 100 tons.
340 Pages. 6x9 Paperback. Illustrated.. $19.95 Code: ATP

HESS AND THE PENGUINS
The Holocaust, Antarctica and the Strange Case of Rudolf Hess
By Joseph P. Farrell

Farrell looks at Hess' mission to make peace with Britain and get rid of Hitler—even a plot to fly Hitler to Britain for capture! How much did Göring and Hitler know of Rudolf Hess' subversive plot, and what happened to Hess? Why was a doppleganger put in Spandau Prison and then "suicided"? Did the British use an early form of mind control on Hess' double? John Foster Dulles of the OSS and CIA suspected as much. Farrell also uncovers the strange death of Admiral Richard Byrd's son in 1988, about the same time of the death of Hess.

288 Pages. 6x9 Paperback. Illustrated. $19.95. Code: HAPG

HIDDEN FINANCE, ROGUE NETWORKS & SECRET SORCERY
The Fascist International, 9/11, & Penetrated Operations
By Joseph P. Farrell

Farrell investigates the theory that there were not *two* levels to the 9/11 event, but *three*. He says that the twin towers were downed by the force of an exotic energy weapon, one similar to the Tesla energy weapon suggested by Dr. Judy Wood, and ties together the tangled web of missing money, secret technology and involvement of portions of the Saudi royal family. Farrell unravels the many layers behind the 9-11 attack, layers that include the Deutschebank, the Bush family, the German industrialist Carl Duisberg, Saudi Arabian princes and the energy weapons developed by Tesla before WWII.

296 Pages. 6x9 Paperback. Illustrated. $19.95. Code: HFRN

THRICE GREAT HERMETICA & THE JANUS AGE
By Joseph P. Farrell

What do the Fourth Crusade, the exploration of the New World, secret excavations of the Holy Land, and the pontificate of Innocent the Third all have in common? Answer: Venice and the Templars. What do they have in common with Jesus, Gottfried Leibniz, Sir Isaac Newton, Rene Descartes, and the Earl of Oxford? Answer: Egypt and a body of doctrine known as Hermeticism. The hidden role of Venice and Hermeticism reached far and wide, into the plays of Shakespeare (a.k.a. Edward DeVere, Earl of Oxford), into the quest of the three great mathematicians of the Early Enlightenment for a lost form of analysis, and back into the end of the classical era, to little known Egyptian influences at work during the time of Jesus.

354 Pages. 6x9 Paperback. Illustrated. $19.95. Code: TGHJ

ROBOT ZOMBIES
Transhumanism and the Robot Revolution
By Xaviant Haze and Estrella Eguino,

Technology is growing exponentially and the moment when it merges with the human mind, called "The Singularity," is visible in our imminent future. Science and technology are pushing forward, transforming life as we know it—perhaps even giving humans a shot at immortality. Who will benefit from this? This book examines the history and future of robotics, artificial intelligence, zombies and a Transhumanist utopia/dystopia integrating man with machine. Chapters include: Love, Sex and Compassion—Android Style; Humans Aren't Working Like They Used To; Skynet Rises; Blueprints for Transhumans; Kurzweil's Quest; Nanotech Dreams; Zombies Among Us; Cyborgs (Cylons) in Space; Awakening the Human; more. Color Section.

180 Pages. 6x9 Paperback. Illustrated. $16.95. Code: RBTZ

THE GODS IN THE FIELDS

Michael, Mary and Alice-Guardians of Enchanted Britain

By Nigel Graddon

We learn of Britain's special place in the origins of ancient wisdom and of the "Sun-Men" who taught it to a humanity in its infancy. Aspects of these teachings are found all along the St. Michael ley: at Glastonbury, the location of Merlin and Arthur's Avalon; in the design and layout of the extraordinary Somerset Zodiac of which Glastonbury is a major part; in the amazing stone circles and serpentine avenues at Avebury and nearby Silbury Hill: portals to unimaginable worlds of mystery and enchantment; Chapters include: Michael, Mary and Merlin; England's West Country; The Glastonbury Zodiac; Wiltshire; The Gods in the Fields; Michael, Mary and Alice; East of the Line; Table of Michael and Mary Locations; more.

280 Pages. 6x9 Paperback. Illustrated. $19.95. Code: GIF

AXIS OF THE WORLD

The Search for the Oldest American Civilization

by Igor Witkowski

Polish author Witkowski's research reveals remnants of a high civilization that was able to exert its influence on almost the entire planet, and did so with full consciousness. Sites around South America show that this was not just one of the places influenced by this culture, but a place where they built their crowning achievements. Easter Island, in the southeastern Pacific, constitutes one of them. The Rongo-Rongo language that developed there points westward to the Indus Valley. Taken together, the facts presented by Witkowski provide a fresh, new proof that an antediluvian, great civilization flourished several millennia ago.

220 pages. 6x9 Paperback. Illustrated. $18.95. Code: AXOW

LEY LINE & EARTH ENERGIES

An Extraordinary Journey into the Earth's Natural Energy System

by David Cowan & Chris Arnold

The mysterious standing stones, burial grounds and stone circles that lace Europe, the British Isles and other areas have intrigued scientists, writers, artists and travellers through the centuries. How do ley lines work? How did our ancestors use Earth energy to map their sacred sites and burial grounds? How do ghosts and poltergeists interact with Earth energy? How can Earth spirals and black spots affect our health? This exploration shows how natural forces affect our behavior, how they can be used to enhance our health and well being.

368 pages. 6x9 Paperback. Illustrated. $18.95. Code: LLEE

THE MYSTERY OF U-33

By Nigel Graddon

The incredible story of the mystery U-Boats of WWII! Graddon first chronicles the story of the mysterious U-33 that landed in Scotland in 1940 and involved the top-secret Enigma device. He then looks at U-Boat special missions during and after WWII, including U-Boat trips to Antarctica; U-Boats with the curious cargos of liquid mercury; the journey of the Spear of Destiny via U-Boat; the "Black Subs" and more. Chapters and topics include: U-33: The Official Story; The First Questions; Survivors and Deceased; August 1985—the Story Breaks; The Carradale U-boat; The Tale of the Bank Event; In the Wake of U-33; Wrecks; The Greenock Lairs; The Mystery Men; "Brass Bounders at the Admiralty"; Captain's Log; Max Schiller through the Lens; Rudolf Hess; Otto Rahn; U-Boat Special Missions; Neu-Schwabenland; more.

351 Pages. 6x9 Paperback. Illustrated. $19.95. Code: MU33

WATER REALMS
Ancient Water Technologies and Management
By Karen Mutton

From the flushing toilets of ancient Crete to the qanats of Persia, aqueducts of Rome, cascading tank systems of Sri Lanka and the great baths of the Indus Valley to the eel traps of southern Australia, ancients on all continents were managing water in unique ways. Table of Contents includes: The Minoan Waterworks; Case Study—The Tunnel of Eupalinos, Samos; Sicily; Etruscan Waterworks; Aqueducts; Roman Baths; Case Study—Aqua Sulis; Case Study—The Baths of Caracalla; Flood Control Systems; Hydraulic Works in the Provinces; Late Roman & Byzantine Technologies; The Persian Qanat System; Case Study—The Palace of Persepolis; Khmer Empire; Case Study—The Dujiangyan Irrigation System; Hohokam Water Works; Case Study—Teotihuacan; Case Study—The Puquios of Peru; Sardinia Wells; Nymphaea; Celtic Wells; Ancient Fish Traps; more.

254 Pages. 6x9 Paperback. Illustrated. $19.95. Code: WTR

SUNKEN REALMS
A Survey of Underwater Ruins Around the World
By Karen Mutton

Mutton begins by discussing some of the causes for sunken ruins: super-floods; volcanoes; earthquakes at the end of the last great flood; plate tectonics and other theories. From there she launches into a worldwide cataloging of underwater ruins by region. She begins with the many underwater cities in the Mediterranean, and then moves into northern Europe and the North Atlantic. Places covered in this book include: Tartessos; Cadiz; Morocco; Alexandria; Libya; Phoenician and Egyptian sites; Roman era sites; Yarmuta, Lebanon; Cyprus; Malta; Thule & Hyperborea; Canary and Azore Islands; Bahamas; Cuba; Bermuda; Peru; Micronesia; Japan; Indian Ocean; Sri Lanka Land Bridge; Lake Titicaca; and inland lakes in Scotland, Russia, Iran, China, Wisconsin, Florida and more.

282 Pages. 6x9 Paperback. Illustrated. $20.00. Code: SRLM

SUBTERRANEAN REALMS
Subterranean & Rock Cut Structures in Ancient & Medieval Times
By Karen Mutton

Some subterranean structures may have been built for initiation ceremonies or perhaps for acoustic reasons, or both. Mutton discusses such interesting sites as: Derinkuyu, an underground city in Cappadocia, Turkey that housed 20,000 people; Roman catacombs of Domitilla; Palermo Capuchin catacombs; Alexandria catacombs; Paris catacombs; Maltese hypogeum; Rock-cut structures of Petra; Treasury of Atreus, Mycenae; Elephanta Caves, India; Lalibela, Ethiopia; Tarquinia Etruscan necropolis; Hallstatt salt mine; Beijing air raid shelters; Japanese high command Okinawa tunnels; more.

336 Pages. 6x9 Paperback. Illustrated. $19.95. Code: SUBR

SCATTERED SKELETONS IN OUR CLOSET
By Karen Mutton

Mutton gives us the rundown on various hominids, skeletons, anomalous skulls and other "things" from our family tree, including hobbits, pygmies, giants and horned people. Chapters include: Human Origin Theories; Dating Techniques; Evolution Fakes and Mistakes; Creationist Hoaxes and Mistakes; The Tangled Tree of Evolution; The Australopithecine Debate; Homo Habilis; Homo Erectus; Anatomically Modern Humans in Ancient Strata?; Ancient Races of the Americas; The Taklamakan Mummies—Caucasians in Prehistoric China; Strange Skulls; Dolichocephaloids (Coneheads); Pumpkin Head, M Head, Horned Skulls; The Adena Skull; The Boskop Skulls; 'Starchild'; Pygmies of Ancient America; Pedro the Mountain Mummy; Hobbits—Homo Floresiensis; Palau Pygmies; Giants; Goliath; Holocaust of American Giants?; more.

320 Pages. 6x9 Paperback. Illustrated. $18.95. Code: SSIC

ORDER FORM

10% Discount When You Order 3 or More Items!

One Adventure Place
P.O. Box 74
Kempton, Illinois 60946
United States of America
Tel.: 815-253-6390 • Fax: 815-253-6300
Email: auphq@frontiernet.net
http://www.adventuresunlimitedpress.com

ORDERING INSTRUCTIONS

✓ Remit by USD$ Check, Money Order or Credit Card
✓ Visa, Master Card, Discover & AmEx Accepted
✓ Paypal Payments Can Be Made To: info@wexclub.com
✓ Prices May Change Without Notice
✓ 10% Discount for 3 or More Items

SHIPPING CHARGES

United States

✓ POSTAL BOOK RATE
✓ Postal Book Rate { $5.00 First Item, 50¢ Each Additional Item
✓ Priority Mail { $8.50 First Item, $2.00 Each Additional Item
✓ UPS { $9.00 First Item (Minimum 5 Books), $1.50 Each Additional Item
NOTE: UPS Delivery Available to Mainland USA Only

Canada

✓ Postal Air Mail { $19.00 First Item, $3.00 Each Additional Item
✓ Personal Checks or Bank Drafts MUST BE US$ and Drawn on a US Bank
✓ Canadian Postal Money Orders OK
✓ Payment MUST BE US$

All Other Countries

✓ Sorry, No Surface Delivery!
✓ Postal Air Mail { $29.00 First Item, $7.00 Each Additional Item
✓ Checks and Money Orders MUST BE US$ and Drawn on a US Bank or branch.
✓ Paypal Payments Can Be Made in US$ To: info@wexclub.com

SPECIAL NOTES

✓ RETAILERS: Standard Discounts Available
✓ BACKORDERS: We Backorder all Out-of-Stock Items Unless Otherwise Requested
✓ PRO FORMA INVOICES: Available on Request
✓ DVD Return Policy: Replace defective DVDs only

ORDER ONLINE AT: www.adventuresunlimitedpress.com

10% Discount When You Order 3 or More Items!

Please check: ☑

☐ This is my first order ☐ I have ordered before

Name
Address
City
State/Province Postal Code
Country
Phone: Day Evening
Fax Email

Item Code	Item Description	Qty	Total

Please check: ☑

Subtotal ▸
Less Discount-10% for 3 or more items ▸

☐ Postal-Surface
☐ Postal-Air Mail (Priority in USA)
☐ UPS (Mainland USA only)

Balance ▸
Illinois Residents 6.25% Sales Tax ▸
Previous Credit ▸
Shipping ▸
Total (check/MO in USD$ only) ▸

☐ Visa/MasterCard/Discover/American Express

Card Number:
Expiration Date: Security Code:

✓ SEND A CATALOG TO A FRIEND: